Man and Nature in America

Arthur A. Ekirch, JR.

Man and Nature in America

University of Nebraska Press · Lincoln

Library of Congress Catalog Card Number 63–14925
International Standard Book Number 0–8032–5785–6
First Bison Book printing: October 1973
Most recent printing shown by first digit below:
1 2 3 4 5 6 7 8 9 10
Bison Book edition published by arrangement with the author.
Manufactured in the United States of America

To My Children

Cheryl Nancy, Caryl Jocelyn, and Arthur Roger

PREFACE TO THE BISON BOOK EDITION

Almost everyone today is worried about the environment. Over the past decade ecology has gradually become a topic of great popular concern. Aroused citizens vie with scientists and politicians in denouncing the ways in which we exploit the land, contaminate the water, pollute the air, and generally do violence to the historic balance between man and nature. Environmentalists truly have done an impressive job of acquainting us with the seriousness of modern ecological problems. Meanwhile the world's mounting population and complex industrial technology consume ever-increasing amounts of scarce resources. Although no one denies the possible dangers, the complicated state of man's relations with nature seems to defy any easy solution. What is at stake in civilized man's latest dilemma is our entire contemporary way of life with its spiraling appetite for goods and services and its understandable human reluctance to explore the alternative of a lower standard of living.

Ecology is, of course, not exclusively an American interest. The modern world has now become so compressed by advances in the speed of transportation and communication that the concerns of one country affect the others. The Western nations of Europe and America have spread their industrial systems around the globe with the result that the problems of man and nature are now world-wide. Yet in this, as in other international questions, cooperation is difficult to achieve. Nevertheless, although it is neither unique nor the worst among the nations, the United States, as the largest consumer and polluter of the world's resources, is the logical country to set an example of prudential restraint and social economy.

In an important article (*Harper's Magazine*, January 1972), economist Peter F. Drucker has called attention to the high cost of protecting the future environment. The pollutants from human wastes, industrial effluents, and agricultural fertilizers all require scientific answers which may merely increase our dependence upon technology. For example, sewage treatment plants and clear water programs need additional electrical energy which can only be supplied by new power plants which are themselves polluters. Electric automobiles and mass transit, though ecologically desirable substitutes for the internal combustion engine, will probably be delayed indefinitely because of their heavy demand for electrical power.

No matter how desirable and necessary many of the reforms sought by environmentalists may be, they will have to be paid for by reduced corporation profits, higher consumer prices, or increased taxes. While many Americans might benefit from a life style less cluttered by material goods and less dependent on the family automobile, lessened consumption and lower rates of production are bound to affect the national economy and hit hard at the already underemployed minorities of the young, the poor, and the black. In similar fashion, when marginal industrial plants which do not meet environmental pollution standards are shut down, surrounding communities are hurt economically. The crusade to protect the environment may be confronted therefore by the antagonistic social and political need to preserve jobs and homes.

Although it hardly faces the dilemma of certain underdeveloped countries which are being forced to decide whether to continue to use dangerous insecticides or hazard the return of malaria and a high infant death rate, the United States is still likely to be confronted in the 1970s by a series of hard choices. Interested and affected segments of the population may well protest if appropriations for education, slum clearance, health and welfare services are sacrificed to safeguard the environment. America's very success in offering a higher standard of living to greater numbers of its people not only adds to the environmental crisis, but it also renders solutions more difficult. Meanwhile an agrarian way of life, however attractive in theory, can entail unforeseen costs sufficient to defeat its goals. American agriculture, for example, is now almost completely dependent on chemical fertilizers which poison the soil and adulterate the crops. Even more serious is the fact that the residues of these fertilizers become permanent pollutants of the streams and waterways into which they are washed. Yet without the chemicals, agricultural experts contend that American farms could neither maintain their high productivity and profits nor continue to feed the population.

Although the deterioration in man's relations with his habitat has already carried much of the world to a point of no return, some steps can still be taken to restore nature's balance. As a modest beginning, the American government might divert funds hitherto spent on the Vietnam War to environmental needs. Modern war, the most wasteful of man's pursuits, is after all an anachronism in a world in which man's entire environment is imperiled. But unless the world finds a way to apply improved methods of birth control to its mounting population, the old Malthusian checks of war, famine, and disease will continue to mock its efforts to achieve harmonious human or natural relations.

Meanwhile official support is necessary to sustain public enthusiasm. Thus a scientific panel drawn from the National Research Council of the National Academy of Sciences urges in its 1972 report that the United States limit population and conserve resources. Difficulties imposed by the world's growing population, the report notes, pervade all environmental issues. According to the panel, "There can be no effective national or international materials policy that evades the relationship between population, per-capita demand, and environmental impact."

Efforts to resolve our current ecological problems will require the application of the best and most sophisticated scientific knowledge. But these efforts will be wasted unless they substitute for the old idea of the conquest of nature a new ethic in which man lives in harmony with his environment. The history of man's relationship with nature is necessary to an understanding of modern ecology. It also serves as a challenge to all of us.

March, 1973

PREFACE

I have written *Man and Nature in America* in the hope of providing today's readers with some historical perspective on the problem implied in this title. Since I am an historian and not a scientist, I cannot claim expert knowledge or original wisdom in all the fields I have surveyed. But I have tried to summarize fairly the representative opinions of leading authorities whose knowledge and wisdom may be greater than mine. I think there can be little question of the importance of our current concerns over man's relationship to his environment. The possibilities of nuclear war and of overpopulation are only two of the most serious and dramatic forms of the historic conflict between philosophies of harmony and balance and of exploitation. I have not attempted, however, to write another history of conservation in the United States, except in the sense of the preservation of both man and nature through the adjustment of the constructive and destructive forces of modern civilization.

In the course of my work on this book, which has stretched over several years, I am unable to list all my debts to friends and colleagues. But I do wish especially to thank Professors William Neumann and Rudolph Von Abele for their criticisms of style and content, and Mrs. Patricia Hudson for her typing of successive drafts of the manuscript.

Also I wish to thank the following publishers for permission to quote from copyrighted materials: Dover Publications, Inc., for *The Brown Decades*, by Lewis Mumford; *The Mississippi Valley Historical Review*, for "Objectives and Methods in Intellectual History," by John C. Greene; Alfred A. Knopf, Inc., for *Democracy in America*, by Alexis de Tocqueville, edited by Phillips Bradley; Holt, Rinehart and Winston, Inc., publishers of the 1960 revised edition, for *Art and Life in America*, by Oliver W. Larkin; The Macmillan Co., for *The American Spirit*, by Charles and Mary Beard; Gifford B. Pinchot, for *The Fight for Conservation*, by Gifford Pinchot; Harcourt, Brace & World, Inc., for *New Frontiers*, by Henry A. Wallace; Longmans, Green and Co. Inc., courtesy of David McKay Co. Inc., for *Within Our Power*, by Raymond B. Fosdick; Harper & Row, Publishers, Inc., for *Russia, the Atom and the West*, by George F. Kennan; William Sloane Associates, for *Grand Canyon*, copyright by Joseph Wood Krutch; The Viking Press Inc., for *Liberty in the Modern State*, by Harold J. Laski.

ARTHUR A. EKIRCH, JR.

Washington, D.C.
January, 1963

CONTENTS

ERRATA

Page 33, line 9: *Read* geography *for* geology
Page 57, line 4: *Read* it *for* its
Page 204, note 7 : *Read* Lewis *for* Louis
Page 222, column 2, line 2: *Read* Lewis *for* Louis
Page 223, column 2, lines 9–10: *Read* 158–60 *for* 158–59

Man and Nature in America

I

INTRODUCTION

Man *and* nature is the basic fundamental fact of history. The relationship is mutual and necessary. Without man nature has no written history. The lower forms of life exerted no important influence on the natural environment, and the story of that interrelationship is not a part of the record of the human past. Until the advent of man it is probable that nature or the natural environment remained relatively stable and unproductive. At least changes before man's coming are either not known to us, or they have not been considered important. Our concern, after all, is with the ways in which historic man has been able to transform his environment.

It appears obvious in our age of nuclear power that modern man, aided by his technological tools, has been gaining in his ability to change the face of the world. Today man, almost everywhere on the earth's surface, exerts a dynamic influence upon his environment. Whether this influence is good or bad does not admit any simple answer. But the question is a deadly serious one. Much of the future of our world may be involved in the problem of whether it is necessary or desirable for mankind to achieve some sort of adjustment to nature. Modern disciples of Malthus, along with other scholars inclined to a pessimistic outlook, warn us that the time is fast approaching when the world's resources and energy will no longer be able to sustain its ever-mounting population. At the same time many scientists and engineers assure us that the technological skills and inventiveness of the individual will continue to give new and improved means of supporting life. Humanist philosophers, in turn, raise the question of whether technology, even if it is sufficient materially, will suffice in terms of man's spiritual and psychological needs. Still other authorities argue that these higher values can be achieved best in eras of relative economic prosperity and material well-being.

Finally, if one contends that balance and harmony are necessary and desirable to avoid the waste of unlimited exploitation of the environment, will it be possible for man to achieve such a state of relative equilibrium? And, if possible, will its realization be attainable only by the most thorough regimentation of life and thought? Will a regimented harmony then prevent the opportunity for any further progress, condemning us all to live in a completely socialized, stabilized world? Determined to avoid the extremes of white or black, will we see all the colors of the rainbow turned into a permanent murky gray? If, on the other hand, we believe that progress is possible only through human ingenuity in devising new methods of exploiting nature, can such a dynamism endure forever? Or will it be progress only toward some cataclysm that will clear the way for a new evolution and repetition of the movement from simplicity to complexity?

Despite the achievement of releasing and controlling nuclear energy, it is by no means certain that man is all powerful. A generation ago Lewis Mumford pointed out that the influence of nature, which is often supposed to be significant only in primitive conditions of life, does not diminish with the progress of civilization. "As a matter of fact," he noted, "the importance of the land increases with civilization: 'Nature' as a system of interests and activities is one of the chief creations of the civilized man." [1] Civilized man must still live within his natural environment, even though the horizons of that environment may be tremendously enlarging in the age of space. And it also seems true that he can ignore or abuse nature only at the price of imperiling the future of his species, bringing down the whole human race in collective mass suicide.

The idea of what is a proper balance between man and nature has changed with the centuries. The Greeks and Romans were preoccupied with environmental influences. Although the Greeks observed that the cutting down of trees led to the loss of soil by erosion, the ancient civilizations largely ignored the significance of man as an agent in altering his environment. Their emphasis was on human society, in its origins and development, rather than on the environmental changes wrought by human culture. The Greeks also suggested the idea of a Golden Age of great soil fertility and worked out a cyclical interpretation of history to explain the decline of this fertility. Later they alluded to the idea of design or purpose in nature.[2]

The Greek concept of an orderly design in nature was revived in the seventeenth and eighteenth centuries. Christians studied and interpreted nature as proof of God's beneficent wisdom, while the scientists and philosophers of the Enlightenment conceived of nature "primarily as a framework of rationally contrived structures fitted as a stage for the activities of intelligent beings. The words *framework*, *structures*, and *stage* all express the dominant sentiment of the stability and permanence of the great features of nature: the fixed stars, the everlasting hills, the eternal seas, the created species. Change was recognized as a real aspect of nature, but a superficial aspect. It contributed variety to nature's panorama, but it could not alter her fundamental structures." Scientists followed Newton in his hypothesis of a fixed universe, and secular as well as religious thought accepted the view that the permanence of nature and some over-all guiding purpose or design went together. Corollary to this general eighteenth-century view of permanence and design in nature was the belief in a perfect harmony or balance of nature. "Whether in the motions of the solar system, or in the mutual preying of animals on each other, or in the cycle of geological processes on the surface of the earth, change in one direction was thought to be compensated by change in another, while nature as a whole remained unchanged." [3]

Although Europeans for some time were able to observe the effects of deforestation and resultant flooding, especially in the Alps, most early students of the natural world believed that man was a weak geologic agent. By the nineteenth century any fears of the deterioration of the earth as a habitable planet were overborne by widespread confidence in the certainty of progress. Progress was interpreted in terms of purposive beneficial control over the environment, and the accidental undesired effects of human action upon the environment were largely ignored, except by a scattering of concerned scientists.

Prior to modern industrialism and the technological and scientific marvels of the twentieth century, nature was usually accepted as the norm in life. Little concern was expressed over man's role in changing his environment. Nature as the norm stood for a world varyingly regarded as good or evil. From one standpoint man had declined as the state of nature yielded to the forces of civilization. Another view saw history as a story of human progress. But, in either case, the record of the changes was interpreted in terms of man's relationship with nature. A certain harmony or balance of man and the forces of nature was assumed. Man either descended from a primitive state of grace, or he ascended the ladder of civilization in accordance with natural laws. Only in comparatively recent times have the new Utopias of technology, industrial progress, and productivity supplanted the older notion of balance and harmony. It is worth noting, however, that as man has grown more confident of his potential ability to control nature, he has also become more and more pessimistic over his own fundamental worth or the essential goodness of human nature.

Thus the discovery of nuclear energy has not been paralleled by the corresponding development of a new human nature. The atomic bomb may have made modern man obsolete, but it has not provided us with his successor. Instead, we live in fear that we may be extinguished in the holocaust of thermonuclear war. Utopian expectations of the possible uses of atomic power have also not been encouraged by the fact that, so far, it has been developed almost exclusively in terms of preparation for war.

Although the scientific discoveries of the last ten or fifteen years have enormously strengthened man's egocentric belief that he can control nature, there is still no proof of the beneficence of such power. Even much of our planned use of the environment may be of limited short-run wisdom. Large dams for the development of hydroelectric power can disrupt agriculture by preventing the normal silting of river valleys. As the silt collects behind the dams, the latter's efficiency for power production is diminished. Thus technological progress may oftentimes invite new problems.

Even more challenging is the realization that today's experimentation in man's use of nature is no longer limited in extent. Modern technology and improved methods of rapid communication insure that new ways of changing the face of the earth are not confined to a single landscape or local area. Technologically less complex societies tend to exploit a single area. But modern society, by pooling and redistributing its products, can exploit the whole world. A simple food-gathering economy will be transformed by the introduction of commercial agriculture. The results may have widespread effects on population levels and on the process of social change. Technological change, of course, is part of the way by which modern man both adapts to and reshapes his environment. But the change can be so violent and pervasive in its geographic scope as to invite a fresh cycle of disasters. Control of the weather, for example, has enormous implications for endangering both private and national interests.

While natural scientists plead for caution and the desirability of man living in some semblance of harmony with nature, physical scientists encourage the growing modern faith in the efficacy of nuclear technology. And, while natural changes in the environment may still take place independent of man, we tend to believe that such changes are of diminishing importance in the face of what seems to be modern man's superhuman power.

The relationship of man and nature is especially interesting in terms of American history. In the comparatively short span of our civilization the cycle of primitivism to industrialism has been compressed and laid bare for study. Less than a century divides the era when America was looked upon as a Garden of Eden or savage wilderness and the time when it took first place as the world's industrial giant. Probably no people have ever so quickly subdued their natural environment, marching across an entire continent and exploiting its resources. Social critics, advocates of conservation, and disciples of Malthus, it is true, have raised their cries of alarm. But for the most part Americans have believed that individual Yankee ingenuity and modern technology will continue to provide new means of development.

A brochure issued recently by one of the largest and most successful American industrial corporations calls technology "the great multiplication table. It is the margin between pint and bushel, between ounce and pound, between dozen and gross. In America, it is the margin between home-spun and nylon, between the pot on the hearth and the eye-level oven, between root cellar and deep freeze, between plow and tractor, between wagon train and jet liner and, because the means must precede the fulfilment, the margin between Lincoln's lonely study by the light of the dying fire and the scholarly haven of the Harvard Yard." Moreover, the company's spokesman asserts in another brochure, "Material progress has not come at the expense of the equally important, non-material values. Rather there has been a parallel growth in all segments of American life." [4]

A thoughtful historian, in arguing against any concepts of geographical or economic determinism or pessimism, defends the contriving human brain as the real pivot of history. "In its planetary relationships, the globe only provides the space setting for man's activities. If he mismanages them he cannot shift the blame to a scapegoat—closed space, nature, God, or Fate, but must assume the responsibility himself for the outcome. In this sense specific events are not inevitable except as men make them so." [5]

This faith in technological progress has a strong hold on the American mind. Much of American history can be regarded as the story of the dynamic release of energy under the favoring auspices of political and economic freedom. It is only in the past seventy-five years or less that Americans have witnessed the antithetical philosophy of social control of the environment with centralized planning and a network of regulatory checks. Although the age of laissez faire may have passed, Americans, peculiarly favored in terms of resources and energy, remain reluctant to accept the idea of control of the environment in the sense of a more thoroughgoing conservation and economizing.

Historically a people of plenty, we are reluctant to practice prudential restraint or to think in terms of possible future scarcity. Thus for most Americans the Age of Anxiety has probably not replaced the Age of Confidence. Man has, after all, survived up to now, though as the contemporary critic Joseph Wood Krutch points out so well, there is much evidence that "man's ingenuity has outrun his intelligence." Though good enough to run his primitive world, he "is not good enough to manage the more complicated and closely integrated world which he is, for the first time, powerful enough to destroy." Logic would seem to indicate that we either reduce the complexity of our world, or that we try to measure up to its technological level by the achievement of a greater wisdom. Instead we seek to attain an ever-higher standard of living, heedless of the fact that, as Mr. Krutch says, "What we ride toward at high speed may not be a more abundant life, but only a more spectacular death." [6] Though science has no free will independent of man, there is the danger that modern man has made its technological wonders, not only a means, but also the end of life. No longer the servant, but the master of man, science has become in addition the possible instrument of his destruction.

Americans find it hard to think in these pessimistic terms. The Second World War, which lowered standards of living and wrought physical havoc in much of the rest of the globe, left the American continents relatively untouched. Surrounded by ever-mounting surpluses of wheat, and never free of concern to find markets abroad for the products of our industries, we consider somewhat casually dire predictions of impending world scarcities. We still think of the problem of plenty or scarcity, and the question of the survival of man in relationship to his environment, in national rather than international terms. Even when realized and accepted, world problems, we feel, cannot be solved in the existing international framework, dominated as it is by strong national rivalries and coldwar animosities. "One World," therefore, though it may be a scientific fact, remains an illusion so far as it concerns the social and political practices of separate nations. Before we can resolve the world-wide problems involved in the relationship between man and nature, we need to know more about national attitudes toward this age-old question. And we need to understand especially the American conception of the balance of man and nature.

II

THE AGRARIAN DREAM

The discovery and settlement of America was a tremendous boon to man's awareness of nature. The American continents were literally and figuratively a New World. At a time when the European environment had lost its pristine bloom, an unspoiled landscape of incredible richness opened up across the Atlantic. Beginning with Columbus, hardly an explorer failed to record his ecstatic comments on the unlimited, natural wealth of the American continents. In a prospectus on the New World the discoverer of America wrote of "fields very green and full of an infinity of fruits as red as scarlet, and everywhere there was the perfume of flowers, and the singing of birds very sweet. In all these regions," Columbus noted, "gold is found among the roots of trees, along the banks, and among the rocks and stones left by torrents." [1] Later, exuberant American patriots agreed with the poet Joel Barlow's *The Vision of Columbus* that the optimism of Columbus had been justified.

America offered varieties in the state of nature to suit every taste. It was a virgin land without technological adornment. Natural resources abounded. This perfect primitivism in environment assumed also an ideal human nature. If man had fallen or lapsed from his primeval simplicity, in America he might recover his original state of innocence. Skeptics, especially those colonists in actual contact with the Indian, often doubted the virtues of the Noble Savage, but the concept had strong romantic appeal. Ultimately and logically it could lead to a kind of animalitarianism. And somber realists also feared that, in any case, the disease of civilization could not be kept from spreading. Eventually, too, America would become a stronghold of antiprimitivism and a center of technology. But an important part of the early American dream, which has never been completely lost, was this belief that here in the New World had been discovered an ideal state of nature capable of breeding an ideal species of man.[2]

Like Columbus, in his notice of fruits and gold, most explorers and settlers who followed him to the New World took a practical rather than an aesthetic view of nature. It was the potential resources, more than the beauty of the primitive wilderness, which roused their imagination. The difficulties in subduing nature and in establishing the first colonies also encouraged a realistic rather than a romantic attitude toward the natural world. Nature was something to be conquered, not passively enjoyed. Yet it was also true that the American continent excited the colonists because its tremendous extent and unparalleled riches made them feel that it could never be conquered, much less exhausted. Thus America, it was believed, would be a perpetual fount or garden, the home of a favored people living in an easy relationship with their environment.

Of the early Americans the New England Puritans had least reason to be pleased with their native clime and landscape, although the resources of the sea were bounteous enough. The enjoyment of nature also formed little part of the stern New England theology. Puritans worshiped a God whose grace was not manifested primarily in nature. Thus they frowned on outdoor sports and hunting for pleasure. But the New England town, with its common or village green surrounded by church and homes, was not unattractive and brought to America something of the inner harmony of the European feudal community.

Colonists more favored in environment than their sisters along the rockbound New England coast generally took a happier view of their natural surroundings. Southerners were encouraged by a warm climate and fertile lands to enjoy outdoor living. Hunting was both a ceremonial cult of gentlemen and the simple pleasure of humbler folk. Planters took pride in their gardens and estates and did not neglect appearances in their strivings for a cash crop. They seemed to follow cheerfully the advice, "Dwell here, live plentifully, and be rich." Slavery was a flaw, but apart from this exploitative arrangement in its labor system, the colonial plantation achieved a remarkable economic harmony. The larger plantations were self-sufficient agrarian units, producing the greater part of their needs. In the long run this system of Southern agriculture revealed certain weaknesses. Soil was exhausted and labor enslaved. But, in its first century, the Southern plantation seemed an almost idyllic case study in the close relationship of man and nature. William Byrd, owner of Westover, set forth the ideal goal of his fellow planters: "A library, a Garden, a Grove, a Purling stream are the Innocent scenes that divert our Leisure." As a later commentator pointed out, "Such simple desires show the modesty of the demands upon nature; yet they also indicate that there were no puritannical restrictions on the use of leisure time or on any pleasant relationship between man and nature." [3]

Although New England towns had their commons and most settlers their gardens, it was Philadelphia under the direction of

William Penn that came closest to a planned and balanced community. Penn resented any wastefulness of natural resources, and he also hoped to keep Philadelphia "a greene country towne which might never be burnt and might always be wholesome." Even before he saw his colony Penn carefully stipulated that an acre of trees should be left for every five acres cleared. Upon arriving in Pennsylvania in 1682, he observed with pleasure the beauty of the woods and flowers. In Philadelphia it was provided that large open squares should remain as parks occupied by only an occasional public building. Even private homes were to be built in the middle of their plots so that there would be room for air and a garden. Penn's foresight helped to make Philadelphia one of the most pleasing of American colonial towns. A haven for the Society of Friends, the Quaker City was also by virtue of William Penn's plans a natural as well as a spiritual refuge.[4]

Early in the eighteenth century Philadelphia became the home of young Benjamin Franklin, who was destined to be colonial America's greatest natural scientist. Franklin had fled the Puritan stronghold of Boston, and American eighteenth-century scientists also rejected the Puritans' sober outlook on nature. Colonial scientists played an important role in revolutionizing the early American attitude toward nature. They helped to persuade their fellow men that nature was more than an obstacle to be conquered. Since the survival of the colonies was no longer threatened, Americans could begin to be more relaxed in their outlook upon the natural world. Americans also caught something of the scientists' eagerness to explore the world of nature. The unspoiled American scene afforded unique opportunities for scientific investigation, and much effort was applied to catalogue the rich flora and fauna of the continent. John Bartram, "the King's botanist," established the first botanical garden in North America near Philadelphia and, together with Franklin and other colonial scientists, was recognized by the Royal Society of London.

Franklin, though his efforts were mainly directed toward practical inventions, also made impressive steps in the direction of gaining a better understanding of the natural world. He played a key role in establishing the American Philosophical Society. He helped to allay time-honored fears of fire and water with his kite experiment and his passion for swimming. "Throughout his entire life," according to one writer, "Franklin attempted to carry out ideas that aided in harmonizing the relation between man and nature and in bringing about improvements in general living conditions." [5]

The scientists' respect for nature was akin to religious worship, and in the eighteenth century the two were, in fact, joined in the religion or philosophy of Deism. Deists, schooled in the world of Newtonian physics, believed in what they called natural religion. God was relegated to a first cause which set the world in being; after which it ran itself according to natural law. Perfection or improvement was to be gained by living in harmony with the natural world and its laws. Deists found their theology in the natural environment and in the lessons taught by nature, and they rejected therefore the discipline of churches and other man-made institutions.

Tom Paine, the celebrated author of *Common Sense* and America's best-known Deist, regarded the harmonious order of nature as evidence of a divine benevolent plan for man. Science, aided by the divine gift of reason, enabled man to understand the laws of nature, and nature, in turn, was a better teacher than books, including even the Bible. Ever fond of science,

Paine, like the Newtonians, thought of nature in the rationalistic terms of his famous deistical book, *The Age of Reason.* As defined by Paine nature was not a romantic state of anarchy, but it meant harmony, law, and order. "When we survey the works of Creation, the revolutions of the planetary system, and the whole economy of what is called nature, which is no other than the laws the Creator has prescribed to matter, we see unerring order and universal harmony reigning throughout the whole. No one part contradicts another. . . . Every thing keeps its appointed time and place. . . . Here then is the standard to which everything must be brought that pretends to be the work or Word of God." [6]

Many of the Founding Fathers were mild Deists, but though Deism waned after the American Revolution, it had done its part in the general eighteenth-century Enlightenment to inspire a new attitude toward nature. After the Revolution the American interest in nature was concentrated increasingly upon the lands to the west, beyond the Appalachian Mountains. With this great undeveloped hinterland it was widely believed that the United States would be able to preserve an ideal balance between the forces of man and the natural world. The new Republic, freed from political bondage, would become the scene of another Golden Age, duplicating the glory of Ancient Greece and Rome.

Benjamin Franklin, for example, although a town dweller and representative of the New World in the metropolis of the Old, believed that agriculture was the basis of national wealth. The American Revolution freed the colonies from paying tribute to English merchants, and the American population could now move westward to compensate for the idleness and extravagance which Franklin saw beginning along the Atlantic seaboard. The danger would come when all the good lands were cultivated, but a nation in which the bulk of the people were sturdy farmers could afford the luxury of a few merchants in its coastal towns. "The great Business of the Continent," Franklin declared with satisfaction in the late 1780's, "is Agriculture. For one Artisan, or Merchant, I suppose, we have at least 100 Farmers, by far the greatest part Cultivators of their own fertile Lands." [7]

Franklin's views were echoed by the Frenchman St. John de Crèvecoeur, who settled down on a farm in New York with an Anglo-American wife after the French and Indian War. Despite the difficulties he encountered during the Revolution, Crèvecoeur retained a lifelong love of America and nature. "How I hate to dwell in those accumulated and crowded cities!" he wrote. "I always delighted to live in the country. Have you never felt at the returning of spring a glow of general pleasure, an indiscernible something that pervades our whole frame, an inward involuntary admiration of everything which surrounds us? 'Tis then the beauties of Nature, everywhere spread, seem to swell every sentiment as she swells every juice." [8] In his celebrated and popular *Letters from an American Farmer,* Crèvecoeur, like Franklin, assumed that the American West would remain indefinitely an agrarian refuge. "Many ages will not see the shores of our great lakes replenished with inland nations, nor the unknown bounds of North America entirely peopled. Who can tell how far it extends? Who can tell the millions of men whom it will feed and contain? for no European foot has as yet travelled half the extent of this mighty continent!" [9]

American society, Crèvecoeur believed, would retain an ideal simplicity and virtue as it spread westward. "Here," he wrote in his famous enquiry "What Is an American?," "are no aristocratical families, no courts, no kings, no bishops, no ecclesiastical dominion, no invisible power giving to a few a very visible one; no great manufacturers employing thousands, no great refinements of luxury. The rich and the poor are not so far removed from each other as they are in Europe. Some few towns excepted, we are all tillers of the earth, from Nova Scotia to West Florida. We are a people of cultivators, scattered over an immense territory. . . . We have no princes, for whom we toil, starve, and bleed; we are the most perfect society now existing in the world." [10]

However popular Crèvecoeur's Arcadian view of America, not everyone accepted the New World as a modern Garden of Eden. By the late eighteenth century, some scholars looked upon America as a barbarous land inhospitable to civilization. In France, Buffon, De Pauw, Raynal, and others pointed to the American Indian's savage state and degeneracy as an example of unfavorable environmental effects. Buffon advanced the opinion that the animals common to both the Old World and the New were inferior and less varied in the latter, while Raynal argued that America had produced no man of genius in a single art or science. Franklin and Jefferson felt that they had to defend their young country, Jefferson penning his *Notes on Virginia* in answer to the French critics. His only original full-length book, the *Notes* was a pioneer scientific work on American geography as well as an interesting expression of Jefferson's faith in the beneficence of the American natural landscape and resources.[11]

In America, Jefferson wrote, "we have an immensity of land courting the industry of the husbandman." Why then turn to manufacturing, especially since "Those who labor in the earth are the chosen people of God, if ever He had a chosen people, whose breasts He has made His peculiar deposit for substantial and genuine virtue." Farming was a natural way of life which prevented a false dependence in the individual and the growth of corruption in the state. "While we have land to labor then, let us never wish to see our citizens occupied at a workbench, or twirling a distaff. . . . The mobs of great cities add just so much to the support of pure government, as sores do to the strength of the human body." [12]

The views that Jefferson expressed in his *Notes* were not just an impractical sentimentalism. Statesman as well as philosopher, he was ever interested in the political aspects of the agrarian ideal. And, better than any of his contemporaries, he was able to translate his philosophy into action. The concept of a harmony between man and nature lay at the heart of his ideal society. Fundamentally an agrarian philosopher, Jefferson believed in the primacy of agriculture as the ideal pursuit for men and nations. The virtue of the United States would be preserved only if it remained true to its agricultural heritage and refrained from undue reliance on trade and manufacturing. His experience abroad taught Jefferson the dangers of large cities and a landless class of factory operatives. "When we get piled upon one another in large cities, as in Europe, we shall become corrupt as in Europe, and go to eating one another as they do there." [13] For America to remain free of such corrupting influences, a proper balance between man and nature was essential. An increasing population without room for expansion represented a threat to the Republic which Jefferson hoped would be averted by the purchase of Louisiana. This tremendous acquisition, doubling the national domain, promised to keep the United States a nation of farmers, while the pleas of Hamilton and the Federalists for the encouragement of manufactures would be minimized.

Such controversial and contradictory policies as the embargo legislation of 1807, with its coercion and centralization, can best be understood in the light of Jefferson's efforts to preserve peace and keep America free from Europe's debilitating influence. Even the turn to manufacturing and internal improvements, which the embargo policy encouraged, was accepted by him if it would keep the United States from undue dependence upon Europe and make possible a self-sufficient domestic economy. Manufacturing therefore was preferable to commerce, and Jefferson bitterly criticized the New Englanders for their hue and cry over their trade and their desire "to convert this great agricultural country into a city of Amsterdam. But," he added, "I trust the good sense of our country will see that its greatest prosperity depends on a due balance between agriculture, manufactures and commerce, and not in this protuberant navigation which has kept us in hot water from the commencement of our government, and is now engaging us in war. That this may be avoided, if it can be done without a surrender of rights, is my sincere prayer." [14] At the close of his presidency, in reviewing the recent policies of the administration, Jefferson wrote that they "have hastened the day when an equilibrium between the occupations of agriculture, manufactures, and commerce, shall simplify our foreign concerns to the exchange only of that surplus which we cannot consume for those articles of reasonable comfort or convenience which we cannot produce." [15]

Although he did not present his views in systematic form, Jefferson was the most important American agrarian. His philosophy has continued to have its nostalgic appeal and groups of followers. In his own time it was given its most detailed expression in the writings of his fellow Virginia planter, John Taylor of Caroline. Like Jefferson, Taylor stressed the rural virtues of an agrarian society. These were in accord with nature and could best be preserved if government refrained from interfering with its citizenry, except to provide individual freedom and equality of opportunity. Property as the natural fruit of a farmer's labor was to be protected, in contrast to property resting on artificial privilege. A society thus based on a natural economy would also, Taylor concluded, be the best balanced one. A policy of laissez faire would encourage a natural harmony of economic interests.[16]

The Taylor-Jefferson dream of balance was hardly realized. For another half century or more the United States remained an agricultural nation, with a succession of frontier farmers moving west and tilling their own acres. But what Jefferson and his fellow agrarians failed to foresee were the dynamic pressures generated by industrialism. Americans were not content to practice subsistence agriculture, nor to enjoy a slow gradual development. Progress was accepted only in terms of a rapid conquest and exploitation of the environment. More men and machines were necessary to fill the factories and build lines of communication across the continent. And so the Jeffersonian dream was short lived. His agrarian philosophy was incompatible not only with Northern industrialism but also with Southern slavery. Yet Jefferson was not a mere dreamer. Unlike European philosophers of primitivism, or American romantics with their worship of the natural beauty of the landscape, Jefferson favored practical compromise and adjustment between man and nature. His type of agrarian society might have made possible an ideal harmony between environment and civilization. But Jefferson's view was too passive and gentle. It credited to environmental factors more than most Americans were willing to concede.

According to one view, "The capital difficulty of the American agrarian tradition is that it accepted the paired but contradictory ideas of nature and civilization as a general principle of historical and social interpretation." [17] Certainly most Americans, rather than trying to live in harmony with nature, believed that they could exploit nature and, with the aid of technology, tremendously multiply its gifts. And so in the course of time, as another historian has pointed out: "American society lost its idyllic qualities. It lost them primarily because of forces that had been inherent in the American character from the beginning. With their drive toward the domination of nature and toward social and economic success, the Americans could not be content with an agrarian way of life. They preferred both the rewards and the hazards of industrial capitalism, and in doing so they sacrificed most of those features of eighteenth-century life which had appeared so admirable." [18]

III

THE ROMANTIC VIEW

The American agrarian dream implied a political philosophy. As long as agrarianism squared fairly well with reality, it could serve as the basis for a Jeffersonian, and even Jacksonian or antebellum Southern, political and economic program. But as the dream seemed to depict the past more than the present or future, it lost its political vitality and was translated into a romantic idealization of nature and protest against technology.

Thus Americans, though refusing to accept nature in terms of the Jeffersonian-Jacksonian agrarian political philosophy, continued to express regret over the passing of the older, rural ways. They looked forward to the progress of civilization with all its technological changes, but they also sentimentalized the Indian and extolled the virtues of the American landscape. Moses Austin, traveling through Kentucky in the winter of 1796–97, found many of the newly constructed homes and public buildings an eyesore. The farther he went the worse things became until he finally lost his patience at Louisville and wrote indignantly that this town "by nature is beautifull but the handy work of Man has insted of improving destroy'd the works of Nature and made it a detestable place." Austin predicted, however, that by the time his son would be of his age "the country

I have passed in a state of Nature will be overspread with Towns and Villages, for it is Not possible a Country which has within itself everything to make its settlers Rich and Happy can remain Unnotic'd by the American people." [1]

Seemingly, most Americans were unworried by the conflict going on between civilization and nature. They lauded Daniel Boone as an advance agent in the conquering march of civilization and also as an unreconstructed lover of nature who could not abide a neighbor within a hundred miles. The biographers of Boone and the American public thus made their hero the symbol of an American empire and of the primitive wilderness, without any awareness of the conflict between the two concepts.[2] It has been suggested that Americans by indulging in this sort of sentimentalism over nature helped soothe their guilty conscience over the rapid material expansion of the nation. Perry Miller, in making this point, has noted the difference "between the American appeal to Romantic Nature and the European. In America, it served not so much for individual or artistic salvation as for an assuaging of national anxiety." [3]

Their treatment of the Indian illustrated the Americans' ambivalent attitude toward nature. However noble a savage, the Indian had to be accommodated to the tempo of civilization. Little attention was paid to the red man's objection to civilization as an artificial contrivance that intruded between man and nature. An educated Indian interviewed by an early American party on its way to Oregon gave a particularly impressive statement of the Indians' concept of the balance of nature. "As soon as you thrust the plowshare under the earth," declared the articulate red man, "it teems with worms and useless weeds. It increases population to an unnatural extent—creates the necessity of penal enactments—spreads over the human face a mask of deception and selfishness—and substitutes villainy, love of wealth and power, and the slaughter of millions for the gratification of some royal cutthroat, in place of the single-minded honesty, the hospitality, honour and the purity of the natural state." [4]

The popular white man's attitude was summed up by President Jackson in his argument for the Indians' removal to the West. "Philanthropy," Jackson asserted, "could not wish to see this continent restored to the condition in which it was found by our forefathers. What good man would prefer a country covered with forests and ranged by a few thousand savages to our extensive Republic, studded with cities, towns, and prosperous farms, embellished with all the improvements which art can devise or industry execute, occupied by more than 12,000,000 happy people, and filled with all the blessings of liberty, civilization, and religion?" [5] In theory the Indians' civilization was possible; in practice he was destroyed. Censured at first for his failure, the red man eventually came to be pitied for his inability to adapt to the white man's ways. Although the passing of the Indians had long been regarded as inevitable, Americans continued to admire the Noble Savage as the symbol of a vanishing age.

Sentimentalism over nature and the Indian because of guilt feelings concerning their exploitation and extermination was a romantic reaction. It was therefore more popular as a theme for artist and author than as a guide to practical political conduct. As Alexis de Tocqueville observed, "In Europe people talk a great deal of the wilds of America, but the Americans themselves never think about them; they are insensible to the wonders of inanimate nature and they may be said not to perceive the mighty forests that surround them till they fall beneath the hatchet. Their eyes are fixed upon another sight: the American people views its own march across these wilds, draining swamps, turning the course of rivers, peopling solitudes, and subduing nature." [6] Though Tocqueville noted that Americans thought of the frontier largely in terms of material progress, like most foreign travelers he, too, was much impressed by this aspect of the American West. European observers, as well as American pioneers, believed that the West would be a guarantee of free institutions and of prosperity in the future. Among the principal causes for the success of the democratic republic in the United States, Tocqueville wrote, was "the nature of the territory that the Americans inhabit. . . . In what part of human history can be found anything similar to what is passing before our eyes in North America? . . . Everything is extraordinary in America, the social condition of the inhabitants as well as the laws; but the soil upon which these institutions are founded is more extraordinary than all the rest. . . . That continent still presents, as it did in the primeval time, rivers that rise from never failing sources, green and moist solitudes, and limitless fields which the plowshare of the husbandman has never turned. In this state it is offered to man, not barbarous, ignorant, and isolated, as he was in the early ages, but already in possession of the most important secrets of nature, united to his fellow men, and instructed by the experience of fifty centuries." [7]

In the early years of the nineteenth century, American writers and artists, not surprisingly, were more inclined than the Western frontiersman to entertain a romantic view of the world of nature. Abroad, Wordsworth and the English Lake Poets were at the height of their fame, and they exercised a stimulating influence upon their American contemporaries.[8] But the love of nature was also a matter of national pride, and American writers and artists urged their fellows to visit and appreciate local, American beauty spots. Washington Irving, the first native writer to achieve a reputation abroad, in his Knickerbocker history and Sleepy Hollow tales created a pleasant picture of nature in the Hudson River Valley of New York. In contrast to the bustling citizenry of the nineteenth century, the placid Dutchmen in New Netherlands seemed in Irving's pages to live in contented adjustment to their natural environment. William Cullen Bryant, America's first real poet of nature, in "Thanatopsis" described the effect of the natural world upon the individual.

> To him who in the love of Nature holds
> Communion with her visible forms, she speaks
> A various language.

But "Thanatopsis" also concluded in the fashionable mode of a gloomy sentimentalism, with the individual returned at death to his eternal resting place in nature.

The poets' interest in nature and the American landscape was paralleled by the romantic school of early American painters. On a professional level, American landscape art came of age in the 1820's. "From the topographical it had moved to the lyrical, the grand, the allegorical; and nature in this decade could mean a variety of things to a variety of people. Both national pride and response to the marvelous welcomed pictures in which the wonders of nature were celebrated, while the writers reminded one that a pushing civilization had not yet destroyed those wonders. John Vanderlyn was one of many, though perhaps the first, to sketch 'sublime' Niagara." [9] Americans, however, were not ready to surrender themselves completely to the loveliness of nature, and American landscapes frequently conveyed a message with strong moral overtones. This was true, for example, of the work of Washington Allston.

The most important and sympathetic of American landscape painters in their feeling for nature was the so-called Hudson River School of the 1830's and '40's. Leader of the group was Thomas Cole, whose pictures, Bryant said, "carried the eye over scenes of wild grandeur peculiar to our country, over our aerial mountain tops with their mighty growth of forests never touched by the axe, along the banks of streams never deformed by culture and into the depths of skies bright with the hues of our own climate." [10] Before the Hudson River School, landscapes had never paid, and artists had of necessity to turn to portraiture. But the dwindling frontier and vanishing rural life of the East made Americans eager to buy paintings of the pastoral countryside or of the more rugged West.

The Hudson River was a popular choice for the early landscapist. Irving had already celebrated its charms in literature, and the Erie Canal had increased traffic up and down the river. Many a traveler therefore was familiar with every turn and bend of the stream and the pleasing views of the Catskill Mountains to the west. It was also generally a region that deserved the term "picturesque." In contrast to other rivers and mountains, the Hudson and the Catskills were gentle but strong. The river made its way serenely, unbroken by falls, for over one hundred and fifty miles. The neighboring mountains were not jagged and rough but smooth enough to afford broad vistas over rolling country, in which farms were beginning to dot the forest. Thus the Catskills and the Hudson had undoubted appeal to a romantic generation of artists and writers. While Thomas Cole and the Eastern landscape artists painted these quiet places, their more venturesome fellows turned to the West.

There in the 1830's George Catlin began his famous Indian scenes, while others explored the rugged beauty of the Rocky Mountains.

The inspiration that American painters found in the Catskill Mountains and in the vanishing Indian also stimulated the pen of James Fenimore Cooper. In his novels he deplored the destruction of the natural beauty of this same mountain area beloved by the romantic painters. Cooper particularly feared the wanton cutting down of the forests. Favoring an intelligent use of resources, he did not wish to forestall all civilization or technology, but offered as an example of a modern village "not one of those places that shoot up in a day, under the unnatural effects of speculation, or which, favored by peculiar advantages in the way of trade, becomes a precocious city while the stumps still stand in the streets; but a sober country town, that has advanced steadily *pari passu* with the surrounding country, and offers a fair specimen of the more regular advancement of the whole nation in its progress towards civilization." [11]

Cooper's famous character Leatherstocking was a child of the uninhibited wildness of the frontier, in conflict with the more settled, civilized society illustrated in Cooper's own family estate at Cooperstown. The latter represented the kind of compromise between man and nature which Cooper, especially in his later years, came to admire. More important to him than the enjoyment of nature aesthetically was the problem of balancing the household of nature. In *The Pioneers* Cooper developed the point that man should preserve the beauty and resources of nature. Both Leatherstocking and the Judge complained of the destruction of wild life and forest. Noting the use of maple sugar at his table and the continued cutting down of trees for firewood, the Judge observed: "If we go on in this way, twenty years hence we shall want fuel." [12]

Nathaniel Hawthorne, Cooper's younger contemporary, shared his senior's anxiety over a too rapid progress of civilization. Scorning the uncritical, mass American faith in a material progress which neglected important individual and spiritual values, Hawthorne satirized the popular adulation of change without improvement. In *Main Street*, one of his shorter pieces written in the 1840's, Hawthorne imaginatively recreated the history of Salem, Massachusetts. As the panorama of the past life of Salem was unfolded with its mixture of Puritan worthies, saints, witches, and criminals, he commented on the changes that had occurred in the natural landscape, particularly the destruction of the Salem forests. A reborn Indian medicine man who affrighted the whites would himself have been more frightened, Hawthorne wrote, if "he could catch a prophetic glimpse of the noon-day marvels which the white man is destined to achieve." [13]

As Main Street overcame the rural countryside, the romantic view of nature was tamed. Leatherstocking's longing for the frontier wilderness had to be satisfied on the Far Western prairies, while his compatriots in the East turned to the pen or brush to preserve a picture of the wildness of nature. In more realistic fashion, a few far-sighted individuals urged the establishment of parks to keep a bit of nature in the city. Even before the Civil War it was apparent that many of the larger American cities were already dirty, unattractive, overcrowded, and unhealthy. The natural beauty of their river banks and harbors was being sacrificed to the practical needs of trade, and factory smoke filled the air.

The first efforts at city planning were a mixture of the romantic and the practical. In Philadelphia William Penn's original idea for a planned city was neglected in favor of commercial convenience, and the Quaker City lost the pleasing aspect of its colonial days. But in 1828 Philadelphia acquired the first land for what later became the four thousand acres of Fairmont Park. Boston meanwhile landscaped its Common, and in New York City the Battery and Bowling Green were still pleasant parks in the 1830's. In Washington the magnificent blueprint for the nation's capital laid out by L'Enfant was ignored, but the city on the Potomac was still small and almost rural in atmosphere. It was also little affected by trade and industry.

The organized movement for the modern-type city park seems to have had its inception in the idea of making American burial grounds into scenic cemeteries. The man who probably originated this idea was a Boston physician, Dr. Jacob Bigelow, who was also author of a pioneer treatise on technology. Concerned over both the hygienic and aesthetic state of the usual burying places, Bigelow advised placing cemeteries outside the city in rural scenic surroundings. In 1831 Mount Auburn Cemetery, situated four miles from Boston, became "the first example in modern times of so large a tract of ground being selected for its natural beauties and submitted to the processes of landscape gardening to prepare for the reception of the dead." [14] Cemeteries soon became popular visiting places. According to Andrew Jackson Downing, the landscape architect and writer, "People seem to go there to enjoy themselves, and not to indulge in any serious recollections or regrets. . . . Indeed," he noted, "these cemeteries are the only places in the country that can give an untravelled American any idea of the beauty of many of the public parks and gardens abroad." [15]

From landscaped grounds for the deceased, the logical next idea, of course, was parks for the living. Whether or not the transition was so direct, there was point to Downing's query: "If 30,000 persons visit a cemetery in a single season, would not a large public garden be equally a matter of curious investigation?" [16] The most ambitious attempt to preserve something of the beauty of nature in the midst of a large city was the creation of Central Park in New York. Early in the 1840's William Cullen Bryant, the nature poet turned city newspaper editor, tried to interest the people of New York in the idea of a large public park. Bryant was joined in his efforts by Downing who, in the pages of his magazine, *The Horticulturist*, carried on a campaign to acquaint the public with the advantages of parks and gardens. America, Downing pointed out, was far behind European cities in providing attractive and spacious parks for its citizens. Although the first appropriation was made in 1851, Central Park in New York City did not realize the hopes of its advocates until Frederick Law Olmsted, a friend and pupil of Downing's, was installed as superintendent of the project in the late 1850's.[17]

The public park gained support only when it became apparent that the tremendous growth of American cities would soon condemn whole generations to live their lives largely untouched by natural beauty. Even private parks in the form of the country estates of a European landowning class were lacking in America, at least in the North. In contrast to Southern plantations, the town houses of Northern factory owners were almost as devoid of natural beauty as their mills. Obliteration of the American landscape had long been a national passion, and the triumph of steam power threatened still greater damage to the entire countryside. The engineering of the former wood-and-water stage of industrial economy, marked by water wheel and local mill, dirt road, and canal and river transportation, had often provided an attractive landscape. But with the coming of the railroad, as Lewis Mumford pointed out, "the whole picture altered. Railroad cuts were made with no thought of their effect on the landscape; the use of soft coal as a fuel cast a pall over the whole landscape and covered the cities into which the railroads nosed with grime. . . . Blight and waste came in with the boasted prosperities of the early industrial period; and at first the advantages and the defilements were so closely associated that people even prided themselves on the smoke of the thriving town." [18]

Despite the pull of progress it was fortunate that a few practical romanticists foresaw the need to create in American cities small oases of beauty and examples of relative balance between the forces of man and nature. At the same time parks also came to be viewed as a necessity from the standpoint of public health and recreation. They afforded areas for outdoor sports and, for those who could not travel the increasing distances to the country, city parks became the only open spaces available. Immigrants, pouring into Atlantic coast ports, not only increased urban congestion, but they also brought with them firsthand familiarity with the long-established parks of European cities. Thus their influence encouraged American interest.

Parks, though providing windows to nature, could not preserve all its wilder aspects. The West therefore was the best escape for the true lovers of the primitive. For those confined to an urban environment, science paradoxically provided some relief. By their writings and in their museums of natural history, scientists aroused people to a new appreciation of the beauties and wonders of nature. Especially noteworthy was Louis Agassiz, the celebrated Swiss geologist, who came to America in 1846 and began a long and distinguished career teaching at Harvard in 1848. Agassiz imparted to his students, not only careful scientific methods, but also a love of nature. By public lectures and by encouraging the establishment of scientific museums, he was able to popularize science and to carry his ideas to a larger audience than his college classroom.

Although Agassiz was exceptional in transferring his life work from Europe to the United States, other scientists of equal standing shared his admiration for the natural richness of the American environment. One of the first and most influential naturalist explorers of the New World was the vastly learned German scholar Alexander von Humboldt. In *Cosmos*, his great work on physical geology, published originally in five volumes from 1845 to 1862, Humboldt sought to formulate the known facts about the universe into a uniform conception of nature. Humboldt carried over into the nineteenth century the optimistic eighteenth-century faith in the benevolence of nature, and this optimism was reinforced by his extensive travels in America. Early in the nineteenth century he visited the United States and spent some time with Thomas Jefferson. Later, his enthusiasm over the resources and prospects of the young Republic was returned by the favorable reception in the United States of his *Cosmos*.

A successor to Humboldt in American esteem was Arnold Guyot, a Swiss scientist and subsequently professor of geology at Princeton, who came to the United States in 1848 to deliver a series of lectures which he published under the title of *Earth and Man*. Guyot's book, which enjoyed a great vogue, stressed the considerable geographic advantages enjoyed by the United States. The New World and its great West, he believed, would be the scene of the future progress of civilization. Less cautious than Guyot and Humboldt in asserting the need of a proper balance between man and nature, American advocates of manifest destiny used their optimistic theories to buttress their own nationalistic persuasions.

William Gilpin, noted by a recent scholar as the most ambitious student of the American West in the Civil War generation, was one of those who applied the theses of Humboldt and Guyot to his native region. Gilpin eventually reached the point in his interpretation of American manifest destiny where it rested on a geographical determination, almost independent of man. According to Gilpin, following Humboldt's conception of the wholeness of nature, the American continent was a supreme and unbreakable arrangement of mountain, plains, and rivers. This order would eventually accomplish the supremacy of the civilization of the United States, elevating it to a pinnacle above all other nations.[19]

Gilpin was an extreme environmentalist in his conception of American manifest destiny, and he expressed well the varied aspects of the romantic view of man and nature in the first half of the nineteenth century. Romantic notions of nature cloaked realistic national ambitions for western territorial expansion. But manifest destiny as it carried civilization to the Pacific also helped maintain some of the agrarian virtues. It preserved for another generation or two the rural primitive life of the frontier and postponed the impact of technological progress.

IV

TECHNOLOGY AND PROGRESS

Both the American agrarian dream of the Jeffersonian era and the romanticism of the early nineteenth century ran up against the hard realities of technological progress. Though the final triumph of industrialism would be deferred until after the Civil War, the antebellum years were already a period of great material and technological expansion. During these years the acquisition of new raw materials and markets was accompanied by an increasing population. Expansion westward to the Pacific uncovered areas rich in natural resources awaiting development by a growing people. To provide the necessary labor force, the large natural increase at home was augmented by the vast numbers of immigrants coming from Europe. Discouraged by the toils and privations of life in the Old World, they came to America in the hope of sharing its progress.

During the Middle Period the problem of how the land should be used and of whether population growth should be encouraged provoked an endless diversity of opinion. Neither of the two major political parties that sprang from the Jeffersonian Republicans shared Jefferson's view of the harmony of man and nature. The National Republicans or Whigs espoused a program of internal improvements designed to encourage the fullest possible utilization of the natural environment, while the Jacksonian Democrats favored western expansion and a cheap land policy.

Whig Party leaders urged that the Federal government encourage production through a system of tariffs, land subsidies, and improved transportation facilities. President John Quincy Adams was an early and important advocate of such a program. Adams believed that "a progressive improvement in the condition of man is apparently the purpose of a superintending Providence." He wished to see the American people improve this heritage, and he was not opposed to the economic intervention of the Federal government. In his first annual message to Congress in 1825, he declared: "The great object of the institution of civil government is the improvement of the conditions of those who are parties to the social compact. . . . Roads and canals, by multiplying and facilitating the communications and intercourse between distant regions and multitudes of men, are among the most important means of improvement." [1]

The program of internal improvements suggested by President Adams was presented in its most enduring political form in Henry Clay's "American System." The American System was an effort to balance and harmonize the economic interests of the different sections, each one receiving its due share of the resources of the country. To achieve this miracle of planning, Clay offered the nation a program of tariffs, subsidies, and public works or internal improvements. Eastern manufacturers would be protected by a tariff wall as would certain agricultural products faced with foreign competition. The domestic market would be encouraged and expanded by the construction of improved means of internal transportation. Thus the sections would be better able to sell to each other, and American prosperity would rest squarely on home consumption. Manufacturer, workingman, and farmer would be relieved of economic uncertainty. In practice, however, Southern planters had to send their cotton abroad and wished to buy in the world market. And labor seemed to vote for the Democrats, the political opponents of Henry Clay. Thus the American System was never able to gain the national political support necessary to put its author in the White House.

Among those who provided intellectual backing for Clay's American System, none was more interesting or thoroughgoing in his ideas than the economist Henry C. Carey. As Charles Beard later recognized, Carey anticipated in many ways Beard's own version of Continentalism, or economic self-sufficiency and independence for the American continents. Carey was one of the first economic theorists to grapple with the question of what kind of an economic policy would be best suited to the United States. Should the country remain in an agrarian state, producing raw materials for the industrial nations of Europe, or should it attempt a more diversified way of life, seeking to achieve a harmony of agricultural and industrial interests?

Carey was the son of Matthew Carey, an Irish refugee who became a highly successful publisher in Baltimore and Philadelphia. After the War of 1812 the elder Carey urged a policy of economic nationalism for his adopted land. As a son of Ireland, he feared British commercial policy and hoped that a protective tariff would preserve American economic independence. Called a co-author with Henry Clay of the American System, Carey saw his economic ideas developed in the theories of his son Henry. The younger Carey at first followed the precepts of the British classical economists, and it was not until 1845 that he broke with free trade, though still adhering to the general principles of individual initiative and private enterprise.

Carey particularly dissented from the pessimistic theories of the English classical economists Ricardo and Malthus. Ricardo's law of rent asserted that an increasing population forced the cultivation of poorer and poorer soils so that the return to the landlord in the form of rent increased, while the return to labor as wages declined. In Carey's words: "As a necessary consequence of the increasing scarcity of fertile soils, it is held that, with this diminishing return, the land-holder is enabled to take a larger proportion of the proceeds of labor, thus profiting at the cost of the laborer, and by reason of the same causes which tend to the gradual subjugation of the latter to the will of his master." [2] Malthus in his law of population argued that the increase in population would always tend to outrun the food supply. Since this was manifestly undesirable, the population was always forcibly reduced by one of the preventive checks of famine, disease, war, or by some form of continence or birth control. According to Carey: "Over-population is held to be a result of a great law of nature, in virtue of which men grow in numbers faster than they can grow the food that is to nourish them." [3]

This economic pessimism and determinism of the British economists Malthus and Ricardo, with its implications of ruthless competition for survival, Carey disowned root and branch. Man, he felt, was not just an economic animal, nor society merely an economic system or mechanism. Instead he posed the possibility of a rational civilization in which man might work out a genuine harmony of interests, conciliating the differing wants of individuals, sections, and even nations. "Civilization," he wrote, "is marked by elevation and equality of physical, moral, intellectual, and political condition, and by the tendency towards union and harmony among men and nations. The highest civilization is marked by the most perfect individuality and the greatest tendency to union, whether of men or of nations." [4] The true mission of the United States, Carey declared, was "to prove that among the people of the world, whether agriculturists, manufacturers, or merchants, there is a perfect harmony of interests." [5]

Carey deplored what he, and many of his fellow Americans, felt were the wretched economic conditions of the English masses. The traditional hold of the English landlord class and the rise of the new industrial aristocracy, he believed, were responsible for the plight of English labor. With its abundant natural resources, and under the kind of an economic program that he proposed, Carey was confident that the United States could avoid Britain's unfortunate example. First, a protective tariff was necessary to free the United States from dependence on English manufactures and to encourage the development of a diversified industry at home. True independence from Great Britain would also free the United States from entanglement in British colonial and imperial policies and from British wars. Two systems were before the world: the one exemplified in the policies of Britain; the other in those he hoped to see practiced by the United States. "One looks to pauperism, ignorance, depopulation, and barbarism; the other to increasing wealth, comfort, intelligence, combination of action, and civilization. One looks toward universal war; the other towards universal peace. One is the English system; the other we may be proud to call the American system, for it is the only one ever devised the tendency of which was that of elevating while equalizing the condition of man throughout the world." [6]

Although a critic of British free trade and of a gold standard for the currency, which he regarded as the twin bases of British world dominion, Carey was not a dedicated protectionist or apologist for American manufacturers. Agriculture still took predominance over industry, and he opposed any concentration of wealth or creation of a class of Old World workingmen. He rejected also a too rapid development of the American West because of the waste of resources and manpower that would be involved and because he felt it was more necessary to build up the East in competition with England. A foe of the political centralization of power, Carey nevertheless favored a large measure of concentration in the population and economy of a country. Thus his plea was for a more intensive use of resources nearer Eastern seaboard markets. Though Carey did not stress explicitly the idea of a balance between man and nature, his ideas in some ways made him an early American conservationist and social planner. Most especially, he urged the need for a judicious blending of the varied economic interests of the nation in such a way that all would profit to their mutual advantage.

In regard to the relationship of man and nature, the Jacksonian Democrats did not differ significantly from their Whig rivals. Although not favoring internal improvements at Federal expense, they pushed through inflationary policies in regard to cheap land and easy money which encouraged the rapid settlement of the West. The party of the common man, if this was understood to include those who might also be termed small capitalists, the Jacksonians stood for rapid exploitation of the environment. Only in an expanding society could the citizen quickly change his status. If industrialism threatened the worker with the slums and degradation already apparent in England, it also offered opportunities along with risks. Accordingly neither labor, nor the critics of society allied with the workingmen, attacked the technology of industrialism. They and the

Jacksonians complained rather of the lack of equality of opportunity. Machinery in the right social system might be a blessing, not a curse, and an ever-greater production could be a force for social and economic equality.

The economic problems associated with the industrial revolution in England made an especially strong impression upon the American mind. The plight of the English laboring poor and the ugly industrial and urban slum areas in Britain were the subject of much American comment. In the 1820's and 1830's American writers were already beginning to wonder whether labor-saving machinery might not result in surplus production and idleness on the part of labor.[7] The first cotton mills in Massachusetts, in order to allay possible public criticism over the degradation of American labor, evolved their paternalistic Lowell System. The farm girls hired by the factories lived in company boarding houses and were subjected to the disciplinary concern of the owners, who also took pride in the appearance and cultural interests of the girls. Although leisure hours were few, the girls' private lives were rigidly supervised, somewhat in the fashion of students in a women's college dormitory.

In labor circles, or more accurately among the intellectuals and Utopian Socialists who acted as labor spokesmen in the 1830's and '40's, some fears were voiced over the effect of technological progress. If technology made possible an equalitarian mass democracy, it also threatened the integrity of the individual workingman. In many ways, therefore, the Utopian Socialist schemes of Robert Owen and others were efforts to utilize the better side of technology without incurring its accompanying social problems. Robert Dale Owen, Robert's son, declared in 1830: "I see that the immense modern powers of production *might be* a blessing, but that they *are* a curse. I see that machinery, instead of aiding the laborer, is brought into the market against him; and that it thus reduces his wages and injures his situation." [8] The younger Owen was accused by a correspondent in the *Working Man's Advocate* of being an enemy of technological progress. However, he agreed with his critics that the fault lay not in machinery, but in its perversion under the contemporary commercial system.[9]

Closer than the Jacksonians or labor leaders to a theory of balance were the Utopian Socialists in the United States. In communities modeled on the ideas of the English industrialist Robert Owen and the French businessman Charles Fourier, American Utopians attempted to compromise between the forces of nature and modern civilization. Physically, the communities were often located in pleasant rural areas, but near enough to towns and cities to enjoy the advantages of the competitive market until their socialism should be self-sufficing. Within their planned communitarian societies the American Utopian Socialists strove for harmony and balance of man with nature, and of man with his fellow man. Like the labor spokesmen of the Jacksonian era, the Utopians wished to use the powers of science and industry to achieve greater production. The increased material goods would then be shared more equally in the new society of socialism. Except for their hostility to competitive capitalism, they had no real quarrel with the material progress of the early nineteenth century. Charles Lane, the English Utopian Socialist, pointed out that the communities were an extension of the communal property relations of the Indian tribes. He seemed to feel that they afforded a better opportunity to live a natural life than did the usual social arrangements of competitive society.[10]

By the 1840's most of the American socialist communities

were organized on the principles of the French Utopian Charles Fourier. Although Fourier had died earlier, his ideas were carried on by a group of American enthusiasts, notably Albert Brisbane. Fourier's plan of association was based psychologically on the free and full development of human nature. Economically, it proposed to render labor and industry attractive by dividing society into communal groups or phalansteries. According to Brisbane, the first progressive step of the human race had been to develop industry and the arts and sciences; the second was to combine this knowledge and apply it in the superior social system of Fourier's Association. "*The power of Production is unlimited,* and the world may be filled with riches . . . and all may possess and enjoy them abundantly, if *Labor is but rightly organized.* Is not the question worthy of the highest consideration?" [11] Among the partial converts to some of Fourier's ideas was Horace Greeley, who opened up the columns of his *New York Tribune* to Brisbane. Greeley himself pointed out that "Labor working *against* Machinery is inevitably doomed. . . . Labor working *for* Machinery, in which it has no interest, can obtain in the average but a scanty, precarious and diminishing subsistence; while to Labor working *with* Machinery, which it owns and directs, there are ample recompense, steady employment, and the prospect of gradual improvement." [12]

The most interesting and unusual addition to the forty-odd Fourierist societies that existed briefly during the 1840's was the transcendentalist community at Brook Farm, near West Roxbury, Massachusetts. Here George Ripley, the Unitarian minister, had spent a couple of summers enjoying communion with nature and separation from the worldly cares of Boston. From this idyll he conceived the idea of establishing a communal society in which he and his friends might participate. In such ideal natural surroundings it was hoped that both physical and intellectual toil would be at once stimulated and made more pleasurable.[13] Though Emerson and Thoreau refused to join the group formally, there were over a hundred associates, including Nathaniel Hawthorne, who later described the experiment in *The Blithedale Romance*. As Hawthorne perceived, the life of nature and the life of books were not always a fruitful marriage, nor was farming conducive to the placid contemplation of nature. In 1844 Ripley's group departed from their individualism to the extent of turning Brook Farm into a Fourierist community, explaining their hopes in an introductory statement of principles. Association, they declared, was a universal, not a partial, reform under which they were confident "that human life shall yet be developed, not in discord and misery, but in harmony and joy, and that the perfected earth shall at last bear on her bosom a race of men worthy of the name." [14]

Fourierist Socialism, like the early American agrarian dream, was only one facet of the American Utopia. Growing ever stronger in the nineteenth century was the newer faith in science and technology as the way to progress. While the agrarian philosophy of Jefferson and his supporters accepted the concept of a certain balance or harmony between the forces of man and nature, the proponents of technological improvement, including Whigs, Democrats, and even Utopian Socialists, tended to think in terms of a man-made universe. Rather than living in adjustment with the natural world, they all looked for new and better ways to exploit and harness the tremendous resources of nature. Technology, not nature, was becoming the norm, and the power of machinery was replacing the plenitude of nature as the mother of men's hopes for the future. The old American faith in progress, based on the richness of the natural landscape plus the American political system, had to make room for the new belief that science and technology would be the chief foundation stones of future progress. The Age of Nature was yielding to the Age of Science and Invention.

In the nineteenth century civilization and progress were no longer interpreted in terms of the eighteenth century's ideal, harmonious relationship between man and nature. Instead, the nations and peoples regarded as making the greatest advances were those most under the influence of science and technology. Francis Bacon with his inductive system enjoyed a vogue in the United States as the intellectual godfather of modern science because Bacon's emphasis on the experimental method was regarded as more practical than the older deductive, philosophical reasoning. In the same way Americans pointed to the invention of printing and the development of steam power as practical contributions to the welfare of the mass of the people. Printing handed down knowledge from one generation to the next, while the steam engine increased man's physical powers. "This simple, but great machine," the *Scientific American* observed of the steam engine, "has revolutionized the age, and has done more to exalt humanity and benefit the human race, than all the victories of Caesar or the triumphs of Napoleon." [15]

The printing press and the steam engine owed their creation to the genius and talent of an earlier age, but to many Americans their greatest application seemed to be taking place in the democratic society of the New World. Here the practical genius of American inventors contributed numerous additional scientific improvements to transform agriculture, industry, and the home. The poet Walt Whitman urged his readers in 1857 to "Think of the numberless contrivances and inventions for our comfort and luxury which the last half dozen years have brought forth—of our baths and ice houses and ice coolers—of our fly traps and mosquito nets—of house bells and marble mantels and sliding tables—of patent ink-stands and baby jumpers—of serving machines and street-sweeping machines—in a word give but a passing glance at the fat volumes of Patent Office Reports and bless your star that fate has cast your lot in the year of our Lord 1857." [16]

Even the religious press indulged in few pessimistic comments regarding the impact of science and technology. The clergy showed much the same enthusiasm as their parishioners over the advance of technology; they were, however, concerned that science be tempered by moral and social values. Few of their number seemed to believe that the progress of inventions might disrupt any supposed ideal balance of nature. Indeed, it was not at all clear that balance and harmony, rather than exploitation and progress, were part of God's plan for man. Although conservation and control received no particular religious sanction, therefore, the clergy's stress on the non-material aspects of science did interpose a certain note of caution in early industrial America.[17]

In the midst of the public enthusiasm over scientific progress, the most critical note regarding the new god of technology was that sounded by the small group of New England theologians and writers who were called transcendentalists. With their intense individualism and deep love of nature, the transcendentalists, and particularly Emerson and Thoreau, urged more clearly than any of their contemporaries a philosophy of harmony or balance between man and nature.

V

TRANSCENDENTAL HARMONY: EMERSON

For most Americans, by midcentury, technology was the path to progress. Expansion and exploitation were prominent national characteristics, and it was only a minority that dared to think in terms of the individual living in harmony or balance with nature. Of the early American observers of man and nature, it was that unique galaxy of New England literati, the transcendentalists, who gave the most profound statement of a philosophy of true harmony and balance. Though but a handful in number, the transcendentalists were influential critics of society. More than any other group of nineteenth-century American thinkers, they were able to divorce themselves from the materialist goals of an ever-greater production and a more systematic use of natural resources. Going further than the romantic poets and painters, who admired nature as the source of beauty, the transcendentalists sought to understand nature in rational as well as aesthetic terms.

Transcendentalism was not a formal philosophy but was rather a faith—one might almost say a religious faith. Emphasizing freedom of will and conscience, it sought to provide an intellectual as well as moral and mystical justification of individualism. An a priori, or deductive, rather than empirical, or inductive, philosophy, transcendentalism was based on the belief that there were certain fundamental truths in the world, not dependent on experience or science, which could not be proved or reasoned out. The transcendentalists' God was a God of love, not hate, who revealed himself in man and nature. Within each individual there was a spark of Divinity. As a guide to conduct, therefore, the individual's own intuition or conscience was superior to his reason or experience. Though each individual man was capable of perfection, Jesus alone had developed fully this potentiality.

In Boston and Concord, Massachusetts, by 1836, a small number of younger men, mostly Unitarian ministers or close sympathizers with that church, came together to discuss ideas to which the world gave the term transcendentalism. Not applying this word to themselves and refusing to formalize their sessions as a club, the group first met in September, 1836, less than two weeks after the publication of Emerson's little book *Nature*. Author of the work that best served as a manifesto of the transcendentalists, Emerson also offered as a challenge to the gathering his remark "that 't was pity that in this Titanic continent, where nature is so grand, genius should be so tame." [1]

In 1836 Emerson, who was the accepted guiding spirit of the circle of transcendental thinkers, was just beginning his life work as a secular preacher and philosophical teacher to the American public. In 1832, a year after his first wife's death, he had resigned his pastorate in Boston and gone to Europe to travel and study and think. Returning home, much recovered in spirits, he delivered his first lectures, married again, and settled down to the important career that had its focus in the serene setting of Concord village.

Nature was the theme of Emerson's early lectures, and nature was also an alternative to formal religion. In the study of nature, in the lives of great men, and in literature and books, he sought substitutes for the teachings of the church. "Where could he better find grounds for his 'First Philosophy' than in nature, already for over a century hailed as the Bible of the Deistic believer? And what faculty could be relied on for the finding with more confidence than the intuition of the individual man, made in the image of the Maker? In the doctrine of correspondence, the assumption of a parallelism between the moral and the natural laws, there was perhaps ground upon which religion and science could meet." [2]

In his early lectures, following his theological and intellectual crisis in 1832, Emerson approached nature through science. "The Uses of Natural History" was the subject of his initial formal address. At the Masonic Temple in Boston in November, 1833, he told his audience that men were designed to be students of nature. Even though the natural occupations of farmer, hunter, shepherd, and fisherman had become less important with the specialized life and labor of modern cities, men were still compelled to acquire considerable knowledge of nature and its properties—water, wood, stone, light, heat, etc. Man's eye was attuned to the beauties of nature, and men still loved the wild. All nature, Emerson urged, was a unity in which man as an observer played his part—observer being fused with the observed. Recalling his own recent visit to the zoological and botanical gardens in Paris, he declared that he was "impressed with a singular conviction that not a form so grotesque, so savage, or so beautiful, but is an expression of something in man the observer. We feel that there is an occult relation between the very worm, the crawling scorpions, and man." [3]

There were numerous reasons for, and advantages to be

gained from, the study of nature, including the "*delight which springs from the contemplation of this truth*, independent of all other considerations." As Emerson pointed out, nature had a salutary effect upon the mind and character of those who cultivated it. "Moreover the state of mind which nature makes indispensable to all such as inquire of her secrets is the best discipline. For she yields no answer to petulance, or dogmatism, or affectation; only to patient docile observation. Whosoever would gain anything of her, must submit to the essential condition of all learning, must go in the spirit of a little child. The naturalist commands nature by obeying her." [4]

It was the teachings of nature, observed at close hand in the country, which gave a superiority of character to rural people. "That flippancy which is apt to be so soon learned in cities is not often found in the country. Nor are men *there* all ground down to the same tame and timid mediocrity which results in cities from the fear of offending and the desire for display." Emerson told his audience that he was confident that "every man who goes by himself into the woods, not at the time occupied by any anxiety of mind, but free to surrender himself to the genius of the place, feels as a boy again without loss of wisdom. In the presence of nature he is a child." Finally, nature was useful in explaining man to himself and in giving him "his true place in the system of being. . . . And this, because the whole of Nature is a metaphor or image of the human Mind. The laws of moral nature answer to those of matter as face to face in a glass. . . . Nature is a language and every new face we learn is a new word." [5]

Deeply concerned though he was over the influence of nature upon man, Emerson was also one of the first American thinkers to consider man's effect upon nature—other than in terms of a crude exploitation and conquest. Speaking on the subject of the relation of man to the globe, Emerson stressed the importance of insuring "that a proportion is faithfully kept, in all the arrangements of nature, between the powers of man and the forces with which he is to contend, for his subsistence." In the contention of man with the animals and with the furies of the elements, he saw "proofs of this adjustment between man and external nature." On the whole, Emerson was optimistic over the way in which man could alter his environment. By digging, draining, ditching, and watering, but most especially by studying nature itself, man kept the world in repair. "But perhaps the most striking effect of the accurate adaptation of man to the globe," he noted, "is found in his love of it. The love of nature—the accord between man and the external world,—what is it but the perception how truly all our senses, and, beyond the senses, the soul, are tuned to the order of things in which we live. . . . I am thrilled with delight by the choral harmony of the whole. Design! It is all design. It is all beauty. It is all astonishment." [6]

The infinite varieties of nature and its ever-changing character were proof to the transcendentalists that it was something more than mere material substance. Moreover, it was the most concrete substantive link with supernatural and spiritual qualities. Intermediary between God and man, nature also carried a portion of the Divinity to each individual. A proper respect for man as well as reverence for nature was a marked feature of the transcendentalists' individualistic and humanistic philosophy. As William Ellery Channing, the distinguished Unitarian minister, and in some ways the intellectual Nestor or godfather of many of the transcendentalists, expressed it: "I do not look on a human being as a machine, made to be kept in action by a foreign force, to accomplish an unvarying succession of motions, to do a fixed amount of work, and then to fall to pieces at death." [7]

The most important statement of the transcendentalist position was provided by Emerson with the publication in 1836 of his book *Nature*. Turning from his first interest in the science of nature, Emerson now integrated nature with his concern with moral philosophy. Nature was the connecting link between God and man. By contemplation and appreciation of nature man could understand life and achieve an original and individual relation to the universe. Civilization catered to external values, while fundamentals were rooted in the individual conscience and in man's bond with the Eternal via nature. Nature, tying man with God or the transcendental Oversoul, was not to be conquered or exploited, but to be understood and appreciated. Since God spoke to man through nature and his conscience, these were better guideposts than man-made rules and governments.

Nature, as Emerson made clear in his early lectures and book, was many things. A physical fact or commodity, it was also beauty and idealism, as well as discipline and language. It conveyed the spirit of the present and a prospect for the future. On the material side, he pointed out: "All science has one aim, namely, to find a theory of nature." But equally important was the spiritual aspect under which "The moral influence of nature upon every individual is that amount of truth which it illustrates to him." [8] Morals, Emerson said, was "the science of the laws of human action as respects right and wrong." To the query, "And what is Right?" he replied: "Right is the conformity to the laws of nature as far as they are known to the human mind." [9]

In drawing up a course of lectures in the fall of 1836, following the publication of *Nature*, Emerson set forth in his outline the propositions that: "1. There is one mind common to all individual men. 2. There is a relation between man and nature, so that whatever is in matter is in mind." He concluded also that "Underneath all appearance, and causing all appearances, are certain eternal laws which we call the Nature of Things." [10]

Emerson appreciated the manifold beauty of nature. "There is more beauty in the morning cloud than the prism can render account of. . . . I never see the dawn break or the sun set," he wrote, "without reflecting, 'What can be conceived so beautiful as actual nature?' " But more important than the aesthetic appeal of nature were the lessons it taught to man. Emerson's major interest was the relationship of man and nature, and the way in which this relationship illustrated a kind of Divine grace. "Natural history by itself," he declared, "has no value; it is like a single sex; but marry it to human history, and it is poetry." [11]

Like his fellow transcendentalists, like advocates of moral reform, and like most sensitive individuals, Emerson was wont at times to inveigh against society and the uncritical admiration of material progress. Head of a large family circle of relatives and friends, and much among the public on his lecture tours, he was never a solitary figure like Thoreau. The individual, he felt, must accept society and its circumstances, and yet must be their master, not their slave. "Solitude is naught and society is naught. Alternate them and the good of each is seen. . . . Undulation, alternation is the condition of progress, of life." If society seemed noxious, nature was the antidote against its baleful influence. "The man comes out of the wrangle of the shop and office, and sees the sky and the woods, and is a man again. He not only quits the cabal, but he finds himself. But how few men see the sky and the woods!" Nature to Emerson was a resource, but it was more than the kind of natural resource that those who thought of it only in physical terms conceived. It was "the beautiful asylum to which we look in all the years of striving and conflict as the assured resource when we shall be driven out of society by ennui or chagrin or persecution or defect of character." [12]

As a young man of twenty Emerson sketched a remarkably mature view of the progress of civilization in relationship to the continuities of nature. Despite the much vaunted progress in the world, he felt that the change from the rude barbarian to the polished modern gentleman was very slight—"at least, what is cast aside is very insignificant. . . . The world changes its masters, but keeps its own identity, and entails upon each new family of the human race, that come to garnish it with names and memorials of themselves,—certain indelible features and unchanging properties. Proud of their birth to a new and brilliant life, each presumptuous generation boasts its dominion over nature; forgetful that these very springing powers within, which nurse this arrogance, are part of the fruits of that Nature, whose secret but omnipotent influence makes them all that they are." Emerson concluded: "The world which they inhabit they call their servant, but it proves the real master. Moulded of its clay, breathing its atmosphere, fed of its elements, they must wear its livery, the livery of corruption and change, and obey the laws which *all* its atoms obey." [13]

Though nature therefore was the real master, it was nevertheless a benevolent despot, dispensing still unexplored riches to its servants or inhabitants. Change, in the sense of gradual evolution, was true of both man and nature and, as Emerson wrote in *Nature*, "It is essential to a true theory of nature and of man, that it should contain somewhat progressive." [14] [*sic*]

In the minds of Emerson and the transcendentalists, the progress of society depended primarily upon the improvement of its individual members. With the publication of his important first series of *Essays* in 1841, Emerson examined in detail this relationship of the individual to society. In the famous piece "Self-Reliance," he took issue with the comfortable American assurance in the inevitable progress of society. Calling society a wave, he maintained that the progress in the arts of civilization had been accompanied by a deterioration in the individual man. "All men plume themselves on the improvement of society, and no man improves," he declared. "Society never advances. It recedes as fast on one side as it gains on the other. . . . For every thing that is given something is taken. Society acquires new arts and loses old instincts." [15]

Scornful of the false deference paid to property and to government, Emerson urged that a greater self-reliance would revolutionize men and institutions. "Nothing can bring you peace but yourself," he wrote. "Nothing can bring you peace but the triumph of principles." Believing that the "infallible index of true progress is found in the tone the man takes," Emerson, against the concept of material progress, placed the idea of the Oversoul with its transcendental implications of truth and beauty. He labeled the life of man "a self-evolving circle," maintaining the view that "this incessant movement and progression which all things partake could never become sensible to us but by contrast to some principle of fixture or stability in the soul: Whilst the eternal generation of circles proceeds, the eternal generator abides." The individual, however, was not bound to the standards of a bygone generation. Limitation was the only sin, or as Emerson expressed it: "There are no fixtures in nature." [16]

Emerson kept clear the distinction between the moral progress of the individual and mere material improvement. True progress came from achieving a harmonious relationship with nature, not from a more efficient exploitation of its beauty and wealth. "This invasion of Nature by Trade," he complained in 1839, "with its Money, its Credit, its Steam, its Railroad, threatens to upset the balance of man, and establish a new, universal Monarchy more tyrannical than Babylon or Rome." Though society prided itself upon each new discovery, it was only the actual inventor or scientist who experienced any real individual transformation. Men generally, despite their boasting, were not improved in their moral being by scientific achievements. It was the inventor alone, Emerson insisted, who "may indeed show his model as sign of a moral force of some sort but not the user." [17]

Parrington calls Emerson the transcendental critic and conscience of America, a voice of positive idealism who salted transcendentalism with a hard core of New England practicality. Combining the best qualities of the Puritan and the Yankee, he represented the Golden Mean of transcendentalism.[18] Emerson enjoyed a wide audience with his lectures and essays. In a world coming daily under the growing influence of technology and industrialism, he was able to find listeners for his philosophy of individualism. But he achieved this success by never straying too far from his public. With all his gibes at progress Emerson could glory in the convenience of the railroad speeding him on his lecture tours, and he could also exclaim "Machinery and Transcendentalism agree well." [19] As a later critic has pointed out, "His ability to keep on friendly terms with his intellectual and social environment and tradition made him a great American mediator; his public accepted from him as gospel what in other tones and idioms its repudiated as heresy and humbug." [20]

The New England transcendentalists were a small minority. But their belief in the concept of a balance or harmony between man and nature stood as the most important social criticism of the popular American philosophy of material progress. Even more than Emerson, though, it was his younger transcendental partner, Henry David Thoreau, who made the most searching criticism of the dominant institutions of his age and who came closest to achieving a philosophy of true harmony.

VI

TRANSCENDENTAL HARMONY: THOREAU

Henry David Thoreau is a classic example of an individual who found inspiration in his own backyard. Unlike Emerson, who visited Europe and lectured throughout the United States, and in contrast to many of the world's great naturalists who explored the continents, Thoreau spent most of his life in his home town of Concord, Massachusetts, a neighbor to Emerson and the little circle of transcendental thinkers. But though the range of his travel was limited, there was no thinness to the keen quality of his observations or to his ability to make narrowness synonymous with intensity and depth. The lessons of nature were to be learned by close observation, and not by superficial travel. Moreover, elaborateness of preparation or arrangement violated Thoreau's basic tenets of simplicity and economy and rendered less possible the achievement of a true harmony of the individual and nature. Restlessness led to exploitation and threatened the balance between the forces of man and nature.

The few books that Thoreau wrote and published during his lifetime, notably *A Week on the Concord and Merrimack Rivers* and *Walden*, are examples of just this sort of leisurely, but acute, understanding and appreciation of the natural world, coupled with quiet indignation at man's abuse of his environment. In these works, and in his extensive *Journal* or diaries, Thoreau spelled out his faith in nature and his hope that its human possibilities could be realized.

For Thoreau, like Emerson, nature was a law and a guide. While man and his institutions were transient, nature provided something permanent to cling to. "I go and come with a strange liberty in Nature. . . . Shall I not have intelligence with the earth? Am I not partly leaves and vegetable mould myself?" [1] Having nature, Thoreau never felt himself an outcast in the way of Melville perhaps. As Sherman Paul points out, "He was certain that nature would sustain him as easily as the stream . . . ," of which he wrote in his first book, *A Week on the Concord and Merrimack Rivers*.[2] Along the Concord and the Merrimack river banks, Thoreau had ample opportunity to see the way in which the nascent New England factories were marring the landscape. The fact that the countryside was still not wholly dominated by an urban industrialism gave some hope to those who wished to avoid the uglier features of the factory system already so apparent in Britain. Thoreau was not a primitive or hermit, but, like Emerson, he wished to see his generation move from an overriding concern with material things to an awareness of spiritual and ideal values. Life was a conflict between industrialism and simplicity, between the exploitation of nature and living in harmony with nature. From industrialism and the factory system Thoreau turned to the transcendental virtues of self-reliance and individualism. By adding his own ascetic brand of a more extreme type of individualism, Thoreau came closer to nature than any of his Concord associates, achieving a philosophy of harmony and adjustment which transcended his own experience and time.

"Men nowhere live as yet a natural life. . . . The poets even have not described it. Man's life must be of equal simplicity and sincerity with nature, and his actions harmonize with her grandeur and beauty." [3] Thoreau thus set down his complaint and offered a remedy. More an individualist than a reformer, he preferred example to agitation and near the end of March, 1845, took himself off to build his cabin and live alone at Walden Pond. He did not wish to play the hermit but to enjoy what he believed was a natural life and describe it to his fellows. Society offered no counsel compared to the lessons of nature. "In society you will not find health, but in nature. Unless our feet at least stood in the midst of nature, all our faces would be pale and livid. Society is always diseased, and the best is the most so." [4]

The record of the year and a half of his experiment that Thoreau set forth in his celebrated volume *Walden, or, Life in the Woods* was the first real attempt by an American to work out a philosophy of man and nature. But before he could describe his own experience, Thoreau had to clear away the superstructure of civilization which overburdened nature. He was unsparing in his criticism of much of what passed for progress. Yet he was not a romantic primitivist who would discard entirely the fruits of civilization. "It would be some advantage," he wrote, "to live a primitive and frontier life, though in the midst of an outward civilization, if only to learn what are the gross necessaries of life and what methods have been taken to obtain them. . . . For the improvements of ages have had but little influence on the essential laws of man's existence: as our skeletons, probably, are not to be distinguished from those of our ancestors." [5]

Thoreau complained of the misuse of the powers of civilization. Civilization, he noted in *Walden*, "is a real advance in the condition of man," but, he added quickly, "only the wise improve their advantages." In the matter of housing, which was his own first practical concern at Walden Pond, civilized man was little better off than the savage. Building materials were more available to modern man than caves to his ancestors, but still Thoreau marveled that it took such a disproportionate amount of time and money to secure shelter. "With a little more wit we might use these materials so as to become richer than the richest now are, and make our civilization a blessing." But civilized man was only "a more experienced and wiser savage," while "The very simplicity and nakedness of man's life in the primitive ages imply this advantage, at least, that they left him still but a sojourner in nature. When he was refreshed with food and sleep, he contemplated his journey again. He dwelt, as it were, in a tent in this world, and was either threading the valleys, or crossing the plains, or climbing the mountain-tops. But lo! men have become the tools of their tools. The man who independently plucked the fruits when he was hungry is become a farmer; and he who stood under a tree for shelter, a housekeeper. We now no longer camp as for a night, but have settled down on earth and forgotten heaven." [6]

Thoreau, of course, had no sympathy for the philosophy that equated progress with the piling up of more luxuries and supposed conveniences. "Most of the luxuries, and many of the so-called comforts of life," he concluded, "are not only not indispensable, but positive hindrances to the elevation of mankind. With respect to luxuries and comforts, the wisest have ever lived a more simple and meagre life than the poor." The improved means of transportation and communication of the modern age resulted undoubtedly in great speed, but frequently the rush was unnecessary because there was "nothing to communicate." In the same way the internal improvements with which the nation was obsessed were superficial and unneeded. Instead of building railroads men should tinker with their lives to improve them. And Thoreau added, "if we stay at home and mind our business, who will want railroads? We do not ride on the railroad; it rides upon us." [7]

In one of his book reviews, the well-known critique of Jacob Etzler's Utopian work *The Paradise within the Reach of All Men,* Thoreau complained that "the chief fault of this book is that it aims to secure the greatest degree of gross comfort and pleasure merely." His quarrel with Etzler was over the mechanistic way he proposed to use nature, bending it to man's will. "How meanly and grossly do we deal with nature! Could we not have a less gross labor? What else do these fine inventions suggest,—magnetism, the daguerreotype, electricity? Can we not do more than cut and trim the forest?—can we not assist in its interior economy, in the circulation of the sap? Now we work superficially and violently. We do not suspect how much might be done to improve our relation to animated nature even; what kindness and refined courtesy there might be." [8]

While Etzler looked forward to the time when man's will would be law to the universe and he would have no more labor than to turn a crank, Thoreau desired to live within the laws of nature and not to transform it. The westward movement of the United States, he feared, would lead only to the further destruction of nature. Would it not be better, he argued, to till the soil of New England and redeem it rather than move to the virgin soil of the West to despoil it? Although Thoreau voiced some regret that men who had once been able to pluck their fruits from the wilderness now had to become cultivators of the soil, he did not despise honest labor or agriculture. The efficient farmer or workingman earned his praise, but he had only scorn for the way in which "Men have become the tools of their tools." [9]

Thoreau's partial primitivism did not call for giving up civilization. It was a plea for simplicity as a means of getting to the essentials of life. And *Walden* was his own experiment to this end. "Our life is frittered away by detail," he complained, and he added the injunction: "Simplicity, simplicity, simplicity!" The pace of life was too swift. "Why should we live with such hurry and waste of life?" he asked. "We are determined to be starved before we are hungry." Foolish wants kept a man both busy and poor. Thus poor John Field, one of Thoreau's nearest neighbors, a hard-working but shiftless man, had to labor "bogging" a meadow for a farmer at the rate of ten dollars an acre, while Thoreau went fishing and appeared a loafer. But, as he explained to Field, he "did not use tea, nor coffee, nor butter, nor milk, nor fresh meat, and so did not have to work to get them; again, as I did not work hard, I did not have to eat hard, and it cost me but a trifle for my food." [10]

Life at Walden was a practical example of simplicity and economy, but Thoreau also savored the experience because it brought him closer to nature and an environment that was still wild and free. "I love the wild not less than the good," he wrote in *Walden,* and he marveled that there remained close to Concord an area so little touched by man. It was ever Thoreau's delight to find for his walks routes unmarred by evidences of civilization. In his *Journal,* he noted: "Nature is very ample and roomy. She has left us plenty of space to move in. As far as I can see from this window, how little life in the landscape! The few birds that flit past do not crowd; they do not fill the valley. The traveller on the highway has no fellow-traveller for miles before or behind him. Nature was generous and not niggardly, certainly." The plenitude of nature, its slow growth, and its perfected whole were all part of the general harmony and balance which Thoreau admired. "I love Nature," he said, "partly *because* she is not man, but a retreat from him. None of his institutions control or pervade her." [11]

The wildness and freedom of nature were best illustrated by the American Indians in their way of life. "The charm of the Indian to me," Thoreau wrote, "is that he stands free and unconstrained in Nature, is her inhabitant and not her guest, and wears her easily and gracefully." But Thoreau also recognized that for "the Indian there is no safety but in the plow." The red man could avoid eventual extermination only by exchanging the role of the hunter for that of the farmer, and by turning from war to diplomacy. In this regard Thoreau observed that the Indian had made greater progress on the whole than the white man. "These savages are equal to us civilized men in their treaties, and, I fear, not essentially worse in their wars." [12]

A major problem in the relationship of man and nature, Thoreau realized, was the incompleteness of man's knowledge of nature. If men would simplify their lives, he felt the laws of the universe would appear less complex. Knowing all the laws of nature, man could infer results from one fact. But having only a partial knowledge, he lacked the essential elements for rational calculation. It was not confusion or irregularity in nature but ignorance in man which prevented his fuller understanding of nature. "Our notions of law and harmony are commonly confined to those instances which we detect; but the harmony which results from a far greater number of seemingly conflicting, but really concurring, laws, which we have not detected, is still more wonderful." Thus there was no chaos in nature.[13]

In his study of the ways of nature Thoreau was particularly intrigued by its interrelationships, the degree of harmony exhibited, and the methods by which nature conserved itself. Though he believed that the world was young and possessed of great wealth, he was indignant that his fellow men callously exploited and upset its natural balance by their wanton destruction of beauty and wildlife as well as physical resources. Thoreau was, in some respects, an early advocate of conservation. He hated to see the farmer cutting and plowing under the pine seedlings from which hardwood forests would in time develop, and he rejoiced in tree planting. Appropriately enough, his greatest interest in conservation was in the possibility of preserving a certain primitiveness in the small local community. A village, even one like Concord, he feared, "would stagnate if it were not for the unexplored forests and meadows which surround it. We need the tonic of wildness. . . . We can never have enough of nature. We must be refreshed by the sight of inexhaustible vigor." [14]

The city Thoreau hated, but the village might be redeemed. Despite all nineteenth-century boasts of the progress of improvement, it was remarkable, he believed, how little the villages and towns did for their own culture and recreation. He was emphatic therefore in urging that each town should have, as well as a schoolhouse, "a park, or rather a primitive forest, of five hundred or a thousand acres, where a stick should never be cut for fuel, a common possession forever, for instruction and recreation. We hear of cow-commons and ministerial lots," he declared, "but we want *men*-commons and lay lots, inalienable forever. Let us keep the New World *new*, preserve all the advantages of living in the country. . . . We boast of our system of education, but why stop at schoolmasters and schoolhouses? We are all schoolmasters, and our schoolhouse is the universe." [15]

Fortunately, the Concord countryside was still rural enough before the Civil War for Thoreau to enjoy both the solitude at Walden and his favorite occupation—walking. "I can easily walk," he noted, "ten, fifteen, twenty, any number of miles, commencing at my own door, without going by any house, without crossing a road except where the fox and the mink do." While he could stand on any of a hundred hills and see the evidences of civilization only in the distance, an Englishman, in contrast, Thoreau observed, was confined to walking in parks and on highways. "I should die from mere nervousness at the thought of such confinement," he added.[16] Roads and highways were meant for horses and business and men in a hurry, while the rivers and woods were for leisure and the study of nature. Every walk to Thoreau was a kind of crusade, but by mid-century, as he lamented, there were few walkers or crusaders. "No wealth can buy the requisite leisure, freedom, and independence which are the capital in this profession." Man's improvements increasingly deformed the landscape, making it "more tame and cheap." "If a man walk in the woods for love of them half of each day, he is in danger of being regarded as a loafer; but if he spends his whole day as a speculator, shearing off those woods and making earth bald before her time, he is esteemed an industrious and enterprising citizen. As if a town had no interest in its forests but to cut them down!" While almost all men were increasingly attracted to society, fewer were being drawn to nature. "Let us improve our opportunities, then," Thoreau ventured, "before the evil days come." [17]

In the 1850's Thoreau could see the odds mounting against the individual and nature. "I hate the present modes of living and getting a living. Farming and shopkeeping and working at a trade or profession are all odious to me. I should relish getting my living in a simple, primitive fashion." Thoreau foresaw unhappily the coming day when huckleberries would have to be purchased from a store instead of being picked from the fields. "The wild fruits of the earth disappear before civilization, or are only to be found in large markets. The whole country becomes, as it were, a town or beaten common, and the fruits left are a few hips and haws." Decrying this sort of specialization of labor, he offered a different sort of division in which everyone could pick his own. "I believe," he asserted, "in the infinite joy and satisfaction of helping myself and others to the extent of my ability. But what is the use trying to live simply, raising what you eat, making what you wear, building what you inhabit, burning what you cut or dig, when those to whom you are allied insanely want and will have a thousand other things." [18]

Thoreau came increasingly to feel "as if the world were on its last legs," that he was living in a tamed and emasculated country like "a tribe of Indians that had lost all its warriors." But he always felt spiritually renewed by a walk in the woods and by the fresh communication with nature that it afforded him. "I suppose that this value, in my case, is equivalent," he wrote, "to what others get by churchgoing and prayer. I come to my solitary woodland walk as the homesick go home." While other men sought society, Thoreau preferred the woods. Not satisfied with ordinary windows, he needed a glade in the forest to see out and around himself. "I must have a true *skylight*," he said.[19]

Nature offered Thoreau a guiding principle, superior to all laws and legislation. The conscience of the individual had its source in nature—its link to the transcendental Oversoul. The author of the essay "Civil Disobedience" did not believe that the individual could resign his conscience to the dictates of the state or to the principle of majority rule. "Any man more right than his neighbors constitutes a majority of one already," he remarked with feeling. "If a man does not keep pace with his companions, perhaps it is because he hears a different drummer. Let him step to the music which he hears, however measured or far away." Legislation had to be right to command obedience, and to be thus right it had to conform to nature. The individual was not to put himself in the attitude of opposition to just government, "if he should chance to meet with such." This possibility, however, was unlikely because nature was as opposed to the state as the individual was to society. Man's primary allegiance, therefore, was to nature and not to the state.[20]

The question, of course, as Sherman Paul has pointed out, was how one could be sure his ideas were based on his conscience and nature, and not on personal whim or idiosyncracy. But, if there was a danger of transcendental madness in transcendental morality, "what saved Thoreau from this predicament was his refusal to use coercion in behalf of his ideas." [21] Thoreau was thus a gentle and mild reformer. Perhaps, because of this, his ideas have endured.

In his life Thoreau was not free of inconsistency. Like Jefferson, he has been criticized for failing to pursue his philosophy in times of stress. In an interesting volume, *After Walden,* Leo Stoller, by an examination of Thoreau's "Changing Views on

Economic Man," has studied the paradoxes or opposites in his thinking. But Thoreau, in his denial of the postulates of material progress, and in his argument for living in harmony with nature, did not waver. And the form of his criticism underscored the whole problem of the possibility of man living in peace with nature. When he died in 1862, the nation, though divided by civil war, was on the eve of another half century of expansion and of the exploitation of its westernmost resources. But as the continent was quickly spanned and brought under cultivation, a scattering of thoughtful observers raised again the questions of balance and harmony that Thoreau had already posed before the Civil War.

VII

GEORGE PERKINS MARSH: PIONEER

In 1862, the same year in which Thoreau died, a scholarly American diplomat completed at Turin, Italy, the manuscript of a work on the interrelations of man and nature. Like the transcendentalists, Marsh was concerned with the human aspects of the natural world and the lessons nature taught. He called his book, which was published in 1864, *Man and Nature; or, Physical Geography as Modified by Human Action.*

From 1861 until his death in 1882, Marsh served as United States Minister to the new kingdom of Italy. A decade earlier he had been Minister to Turkey, and in the 1840's he had been a Congressman from Vermont. Although indifferent to much of the political routine of Washington, he took an active interest in establishing the Smithsonian Institution and in insuring that it would maintain a scholarly purpose. Trained as a lawyer, Marsh's major distinctions, however, were as a scholarly philologist and as a pioneer student of physical geography. A member of one of Vermont's most prominent families, Marsh seemed too restless and intellectually curious to be able to settle down to a comfortable career as a lawyer and politician. His business ventures were almost uniformly unsuccessful, and he turned therefore with relief to the diplomatic appointment which his political connections secured for him. In Italy he enjoyed the stimulating scholarly and social life for which the less than onerous official diplomatic duties gave him ample time. Moreover, he was not tied down as he would have been if he had accepted one of the professorships offered him by several American colleges.[1]

Although Marsh is sometimes considered a forerunner of the conservation movement in the United States, he was actually closer in spirit to the transcendentalists. More interested in ideas and research than in crusading for practical reform legislation, he was not a primitivist or extreme environmentalist but a scholarly humanist who stressed the idea of a proper balance between the forces of man and nature. It was this note which he particularly emphasized in his publications on man's use of the land.

Marsh was a lifelong student of languages, and he possessed a reading or speaking knowledge of a great variety of foreign tongues. His scholarly researches in this field led to publications on philology and also to a well-known address picturesquely entitled *The Goths in New-England.* In this address he called attention to what he believed was the important influence of the Anglo-Saxon, Nordic peoples and institutions upon the United States. Along with this stress on race as a factor in American destiny, Marsh also indicated his later interest in environmental influences. It was the harsh climate of Northern Europe which provided the Goths or Anglo-Saxons with a stimulating homeland. This, as well as racial characteristics, was responsible for their important role in history.[2]

Marsh's *Goths in New-England* was a companion piece to some of the historical works of the period which interpreted

American progress and manifest destiny in terms of the spread of an Anglo-Saxon civilizing mission. He himself believed that American historical writing was too limited and narrow. In a republic history should include the story of the entire population. In other words, it should be social history or the history of civilization. Although holding strong anti-Catholic views along with his Anglo-Saxon interpretation of history, Marsh was not an intolerant chauvinist or narrow patriot. "We are intoxicated with our success, and giddy with the rapidity of our progress," he warned in his discourse *The American Historical School.* The popular belief in progress was often a stimulus to blind and foolish activity. Governments were put under pressure by their own citizens and goaded into unwise policies merely to avoid the stigma of standing still. "Thus government and people are continually acting and re-acting upon each other, and feeding that morbid appetite for novelty and change which threatens to deprive us of all consistence, unity and harmony of national character or institutions." [3]

Environment and inheritance both help to explain Marsh's love of nature. His childhood home in Vermont was beautifully situated at a bend of the river on the lower slopes of Mount Tom, across from the town of Woodstock. The Marsh family's comfortable economic status gave the boy the opportunity to appreciate the countryside without the necessity of the hard work expected of a farm lad. Throughout most of his life Marsh seemed to have had enough leisure to be able always to enjoy nature. And, at the same time, his scientific and scholarly interests kept his love of nature from being merely the romantic satisfaction of idle curiosity. In the numerous excursions which he was able to take while serving abroad, Marsh was ever the observant and interested traveler.

In his letters and despatches he reported fully on what he saw, noting especially the general geography of the area. He collected flora and fauna assiduously, and he was also fascinated by man's works. Thus ancient temples and monuments as well as the world of nature attracted his attention and comment. A humanist like the transcendentalists before him, he was interested in the earth as the home of man and regretted that mankind exhibited so little consideration for its habitat.

Some fifteen years before he set himself to write his *Man and Nature*, Marsh had already explored the general theme of an inner harmony between the two in an interesting address before the Rutland, Vermont, agricultural society. America, Marsh observed, was "the first example of the struggle between civilized man and barbarous uncultivated nature. In all other primitive history, the hero of the scene is a savage, the theatre a wilderness, and the earth has been subdued in the same proportion, and by the same slow process, that man has been civilized. In North America, on the contrary," he pointed out, "the full energies of advanced European civilization, stimulated by its artificial wants and guided by its accumulated intelligence, were brought to bear at once on a desert continent."[4]

Thus in America the process of civilization had been enormously accelerated. The New World enjoyed not only its natural advantages, such as a great variety of plant life, but it had also been able to transplant a number of European grains and vegetables. Although most authorities considered civilization the great enemy of the primitive world of nature, Marsh argued that it was the savage, whose wants could be satisfied only by the exploitation of nature, who was the real enemy. "The arts of the savage are the arts of destruction; he desolates the region he inhabits, his life is a warfare of extermination, a series of hostilities against nature or his fellow man." Civilization, on the contrary, was the mother of peace and renewed the soil by agriculture.[5]

Here Marsh, in a sense, was voicing an ideal for civilization rather than its actual practice, and he hastened to counsel his audience that certain improvements were needed in American agriculture to avoid waste and exploitation. Especially important, he felt, was a better economy in the management of forest lands. "The increasing value of timber and fuel ought to teach us, that trees are no longer what they were in our fathers' time, an incumbrance." The cutting down of the forests led to the erosion of the soil and a loss of humidity. The point of this observation was later reinforced for Marsh by his diplomatic travels around the Mediterranean, where he was able to see lands that had long since lost much of their original forestation.[6]

When President Lincoln appointed Marsh to the diplomatic post in Italy, he gave him the opportunity to settle down and put together the ideas on physical geography that had, for a long time, been one of his major interests. Marsh's earlier diplomatic post in Turkey had placed him in an environment where the effects of deforestation upon the landscape and upon agricultural productivity were obvious. As a scholar he was also able to compare the richness of the ancient civilizations of the Mediterranean world and the impoverished state of much of that same area in his own time. Italy, almost as much as Turkey, afforded a practical example for the theme of his writing. In the Preface to *Man and Nature*, he stated ambitiously that his object was to indicate the character and extent of the changes produced by man in the physical conditions of the globe, the dangers of imprudence, and the necessity of caution; and to suggest the possibilities of restoring disturbed harmonies and of effecting improvements in exhausted areas; and finally to illustrate the possibility of a higher order of man in relation to his environment.[7]

Marsh's thesis was that man disrupts the fundamental harmony or balance of nature. In contrast to the geologists and geographers, and to the environmentalists who stressed the ways in which physical conditions influenced the social life and progress of man, he argued that it was man who exerted a revolutionary effect upon nature. This was both good and bad. Some of man's changes resulted in real progress and improvement along practical lines. But mankind had also destroyed many of the natural advantages it had once enjoyed. This was the case with the Roman Empire. The decline of Rome, Marsh maintained, was caused by a mixture of abuse of the physical environment and bad laws. While nature, undisturbed, fashioned its own balance and redressed geologic convulsion and derangements, man upset this inner harmony. As a familiar example, he pointed to the destruction of birdlife, which encouraged insect growth and thus in turn occasioned tree diseases that further decimated the bird population. "In fine, in countries untrodden by man, the proportions and relative positions of land and water, the atmospheric precipitation and evaporation, the thermometric mean, and the distribution of vegetable and animal life, are subject to change only from geographic influences so slow in their operation that the geographic conditions may be regarded as constant and immutable." [8]

Marsh did not believe that the earth was completely adapted to man's use, and he felt that primitive ways must succumb to human art and technology. "Hence, a certain measure of transformation of terrestrial surface, of suppression of natural, and stimulation of artificially modified productivity becomes necessary. This measure man has unfortunately exceeded." The destructive agency of man seemed to grow with the advance of civilization, "The earth," he asserted, "is fast becoming an unfit home for its noblest inhabitant, and another era of equal human crime and human improvidence, and of like duration with that through which traces of that crime and that improvidence extend, would reduce it to such a condition of impoverished productiveness, of shattered surface, or climatic excess, as to threaten the depravation, barbarism, and perhaps even extinction of the species." [9]

The happier side of this picture, Marsh noted, was the increasing attention paid to the need for conservation of resources. Conservation might also restore some of the damage done by man and nature. For his readers he posed the questions: "how far man can permanently modify and ameliorate those physical conditions of terrestrial surface and climate on which his material welfare depends; how far he can compensate, arrest or retard the deterioration which many of his agricultural and industrial processes tend to produce; and how far he can restore fertility and salubrity to soils which his follies or his crimes have made barren or pestilential?" Among the circumstances that, he felt, gave a particular urgency to his queries was the need for finding new homes for the stream of European immigrants in America. "To supply new hives for the emigrant swarms" was the way he expressed it.[10]

Marsh tried to be happy over the future by reflecting that there was no set limit to the mechanical resourcefulness of man and the possible effects of scientific invention. But a political and moral reformation in the world was needed if technology was to aid conservation. For example, in Holland, since the beginning of the Christian era, he believed that a greater area of land had been lost by erosion of the coastline and drifting sand dunes than had been gained by diking and draining. Yet the soil gained by diking and draining exceeded in value the lands lost. He also pointed out that the costs were no more than those for warships or fortifications, while the new homesites and prosperity, which people had an interest in defending, were "a stronger bulwark against foreign invasion than a ship of the line or a fortress armed with a hundred cannon." [11]

In the United States Marsh was particularly concerned with forest conservation. The desolation already apparent in Europe, he feared, would be the fate of American forests. He did not believe that regulatory measures would be effective unless the state was the owner of the forest lands, although he admitted that tax exemption for such areas might be an alternative. The consequences of forest destruction were climatic changes such as the silting up and flooding of rivers, while "in countries in the temperate zone still chiefly covered with wood, the summers would be cooler, moister, shorter, the winters milder, drier, longer, than in the same regions after the removal of the forest." American life was characterized, he believed, by too much instability and love of change. It was time now, he asserted, to slow down. Enough forests had been felled, and the nation would be wise to establish a fixed ratio of meadow and woodland. Summing up his views, Marsh concluded: "Man has too long forgotten that the earth was given to him for usufruct alone, not for consumption, still less for profligate waste." The work of science in collecting and analyzing data, he regarded as "another step toward the determination of the great question, whether man is of nature or above her." [12]

Marsh's book, though delayed in its publication, attracted

considerable attention and went through several printings. In 1874, ten years after the original edition, he put out a new and revised version entitled *The Earth as Modified by Human Action*. The changes in the new edition were on the whole minor and did not alter his basic thesis. In the same year as its publication Marsh also sent an interesting communication to the United States Commissioner of Agriculture. This paper, called "Irrigation: Its Evils, the Remedies, and the Compensations," was issued as part of the regular report of the Commissioner. It provided a thoughtful analysis of many of the problems which came to the fore a generation later in the conservation movement.

Marsh in his paper on irrigation warned of the danger of embarking with too much enthusiasm upon costly arrangements. European experience demonstrated that irrigation promoted the accumulation of large tracts of land by single owners, while small farmers were dispossessed of their holdings. From an economic point of view, irrigation almost always involved costly construction. "Hence settlers of limited means cannot engage in them, and small land-holding is discouraged." Marsh was exceptionally prescient in foreseeing the American complaints of a later century, or perhaps he had been rendered more aware of the possibilities of overproduction in the United States as a result of the hard times after 1873. In any case, he pointed out that there was a real danger in irrigation in view of the fact that "in some parts of our own country production is now overabundant, that it needs rather to be repressed than enlarged." For example, he noted that the price of corn was not high enough to pay the cost of transporting it to market.[13]

Although Marsh first considered the evils of irrigation, he also proposed remedies for them and noted as well some of the positive advantages to be derived from irrigating the land. If the rapid runoff of mountain waters was dammed near its source and stored for later use, some of the damage caused by deforestation might be repaired as a byproduct of irrigation. In Europe he calculated that the widespread draining of marshy lands had resulted in considerable soil erosion. But conserving and then slowly using the water for irrigation would have the opposite effect of restoring soil fertility. It was a matter of a proper balance or harmony of man and nature. As Marsh wrote: "Draining then deranges the harmony of nature by interfering with her methods of maintaining a regular interchange and circulation of humidity between the atmosphere, the earth, and the sea. Irrigation is in effect a partial return to the economy of our great material parent by regulating that circulation in a manner analogous to her primitive processes." [14]

Familiar with the use of terracing in Europe to prevent soil erosion, Marsh anticipated contour plowing with his recommendation that farmers pursue a process of circling in preparing their soil for seeding. In other agricultural problems, too, he felt that the United States could profit from studying European experience, and he advised the collection and distribution of information on irrigation. Water courses, canals, and reservoirs, he believed, should be owned by the government to facilitate an orderly and equal irrigation of the land. Aware of, and not unsympathetic to, the objection that such powers should not be exercised by a republican government, he, however, felt even more strongly that private citizens could not hold exclusive rights to rivers when water was scarce.[15]

Marsh's report on irrigation has a modern ring. Taken together, his writings on physical geography and the harmony of nature form a link between the transcendentalist philosophy of nature and later conservationist ideology. *Man and Nature* was an important book. Though it emphasized forests rather than minerals or water, it was the first great synthesis treating the various facets of man's use of nature. Almost a century later Marsh enjoyed the belated tribute of having a scholarly symposium devoted to re-examining many of the ideas he had first explored. The collected efforts of these modern scholars assembled at Princeton in 1955 was presented in the noteworthy volume entitled *Man's Role in Changing the Face of the Earth*, which was dedicated to Marsh.[16]

VIII

CONSERVATIONIST IDEOLOGY

Conservation did not become an important American ideology until the 1900's. Only then was it represented by an increasing body of systematic ideas and ardent corps of dedicated disciples. Although there were other influential spokesmen, Theodore Roosevelt and Gifford Pinchot probably did the most to make conservation a popular crusading idea, though, of course, it still meant many different things to various individuals. Moreover, in some of its aspects, it was not new but was as old as American civilization.

During the nineteenth century there had been no lack of a romantic interest in nature, nor of enthusiastic exploration of the American landscape by writers, artists, and travelers. Thoreau and the transcendentalists made nature the center of a philosophy of harmony and balance, and an occasional pioneer like Marsh called attention to the need for conserving natural resources. But America was too much a land of plenty to be worried over alleged or impending scarcities. Instead of diminishing with the nineteenth century, this confident attitude seemed to increase with the settlement of the trans-Mississippi West after the Civil War. Only the so-called closing of the frontier—at least in the sense of the free and easy exploitation of the West—made conservation the serious concern of some Americans.

Nevertheless, for many people, and especially for Westerners, conservation continued to arouse suspicion and hostility. They associated it with the idea of the saving or nonuse of natural resources. Since the prosperity of the West depended on the development of its forests and mineral wealth, Westerners were apt to dismiss conservation as an artificial concept tinged with Eastern romantic and humanitarian notions. Although most conservationist leaders denied that they intended anything more than the curbing of waste and the carefully planned use of natural resources, this view was to some extent a later defense rather than an original argument. Certainly a prominent goal of the conservation movement was the preservation of natural beauty spots and wild life reserves as national parks. And the movement also stressed the necessity of saving forests and mineral resources for the use of future generations. Conservationists were sometimes accused therefore of being nature lovers or socialist planners. But whatever the inner rationale or philosophy of the movement, conservation did seem to point toward the goal of man living in better balance or harmony with his natural environment.

Appreciation of the new wonders of nature as revealed in the Far West played an important part in gaining more general American support for conservation. Along with gold seekers and homesteaders, California in the 1850's was host to a smaller number of visitors who came to admire the gigantic beauty of the Mariposa forest in the Yosemite Valley. After the Civil War the railroads provided easier, swifter access to the West, and the Rockies, Yosemite, Yellowstone, and the Grand Canyon all attracted increasing attention on the part of American literary men and artists. Horace Greeley visited Yosemite in 1859, and after the war Emerson, while in California, received John Muir's enthusiastic invitation to "join me in a month's worship with Nature in the high temples of the great Sierra Crown beyond our holy Yosemite. It will cost you nothing save the time and very little of that for you will be mostly in Eternity." Henry Adams went along on Clarence King's geological exploration of the Fortieth Parallel, and in 1872 William Cullen Bryant edited a two-volume work entitled *Picturesque America*, with illustrations by some of the best American artists. Albert Bierstadt was already famous for his paintings of Far Western scenes, and now in the West for the first time photography began to provide competition for the landscape artist.[1]

Although John Burroughs, Muir's fellow naturalist, had some difficulty in understanding the latter's enthusiastic devotion to the California country, others found in the rugged primeval scenery of the Far West the same kind of a harmony of man with nature that was Burroughs' own main interest. In view of the speed with which the United States had spanned the continent, there was an understandable anxiety to preserve at least some parts of the West from the all-conquering march of civilization. The unique wonders of such sites as Yosemite, Yellowstone, and the Grand Canyon were being portrayed for Easterners by a number of important writers and artists, and in 1872 Congress established Yellowstone as the country's first national park. Other scenic sites were also preserved as national parks, but the program developed slowly until the First World War. There was considerable opposition in Congress to the notion that the government was going into the business of recreation and amusement. At the same time many of the park areas remained inaccessible, and Western tourist travel was confined largely to the Yosemite Valley and Yellowstone National Park. Nevertheless, by 1916, when the National Park Service was created, it had some 37 national parks and monuments placed under its administrative direction.[2]

Although not established as a national monument until 1908, and as a national park until 1919, the Grand Canyon of the Colorado was perhaps the most awesome of the natural wonders of the American Far West. It also became probably the most widely celebrated after the dramatic journey down the gorge led by Major John Wesley Powell in 1869. Until this famous trip Powell had been known as a Civil War hero who courageously continued to serve despite the loss of his right arm in the battle of Shiloh. After the war Powell maintained his boyhood interest in nature by teaching geology. Following summer field trips to the Rockies, he was ready for his epic trip down the Colorado in 1869, making a descent of that part of the river which had never before been successfully navigated. The frightful stories and legends of the river were made real to Powell and his men by their knowledge that after a certain point was reached in the gorge there could be no return, and no escape up the steep-walled banks; the expedition would have to keep going in order to survive, coping with natural hazards of yet unknown size and force. Powell's classic account of the expedition added to the fame of the fearless men who had first braved the rapids of the Colorado, but his major significance for the conservation movement rested on his subsequent career.[3]

In the 1870's Powell continued his exploration of the West, but by 1874 he reported: "There is now left within the territory of the United States no great unexplored region, and exploring expeditions are no longer needed for general purposes." Instead, "It is of the most immediate and pressing importance that a general survey should be made for the purpose of determining the special areas which can . . . be redeemed by irrigation." [4] Four years later he published his noteworthy *Report on the Lands of the Arid Region of the United States.* Observing that two fifths of the United States was arid territory, Powell sketched out the needs of the West for conservation and irrigation.[5] In 1879 Congress appropriated funds for the creation of the United States Geological Survey as an over-all agency to direct the work of the states and Federal agencies in connection with the conservation and protection of the national domain. Powell as director of the survey concerned himself mostly with water-power sites, irrigation projects, and mineral resources. He was a foe of the private water companies which, he believed, wanted to achieve irrigation for farm land without proper conservation practices. Opposed to Social Darwinian notions of an automatic progress by discovering and obeying the laws of nature, Powell declared in an address in 1883 on the Methods of Evolution: "When a man loses faith in himself, and worships nature, and subjects himself to the government of the laws of physical nature, he lapses into stagnation, where mental and moral miasma is bred. All that makes man superior to the beast is the result of his own endeavor to secure happiness. . . . Man lives in the desert by guiding a river thereon and fertilizing the sands with its waters, and the desert is covered with fields and gardens and homes." [6]

Powell's disbelief in an automatic evolutionary progress and his faith in government action and regulation were at variance with the prevailing American mood, but they pointed in the direction which most leaders of the conservation movement followed. Powell's nationalism was moulded by his Civil War army experience and a career as a government scientist. Other leading conservationists were also men of strong nationalistic feelings, and many were familiar with European practice in the regulation of forest preserves and other natural resources. The experience of both Europe and America indicated that the forests which had once covered most of the northern temperate zone were in large part already destroyed. George Perkins Marsh, for example, feared that the forests of the United States would be ravaged in the same fashion as those in Europe. "It is certain," he wrote, "that a desolation, like that which has overwhelmed many once beautiful and fertile regions of Europe, awaits an important part of the territory of the United States . . . unless prompt measures are taken to check the action of destructive causes already in operation." [7]

The preservation of timber resources, which seemed the most pressing conservation problem after the Civil War, also attracted the greatest public attention and sympathy, although the Federal government did not become concerned with forest conservation until late in the nineteenth century. Earlier fears over the possibility of a shortage of ship-construction timber died out with the shift to iron and steel in naval vessels. But by the last third of the century appeals were being made for public action to conserve forest resources. A tree-planting campaign resulted in the national celebration of Arbor Day, and in 1873 Congress passed the Timber Culture Act, making the growing of a certain number of trees a consideration for receiving a quarter section of the public domain. The American Association for the Advancement of Science urged Congress to investigate the condition of the nation's forests, and, in 1875, $2,000 was appropriated to permit the Department of Agriculture to undertake a study of forest conditions in the United States. At this time Secretary of the Interior Carl Schurz, the most prominent of the "forty-eighters" who had come to America from Germany, launched a campaign to stop timber removal from the public domain. In 1877 he recommended that all Federal timber lands be withdrawn from entry under the pre-emption and homestead laws. Although Schurz's efforts met with no immediate success, in the 1880's some of the individual states began to take action in regard to their own forest lands, and public interest continued to grow.[8]

The campaign for forest conservation was given specialized sponsorship in 1882 when the American Forestry Congress was organized. A prominent figure at the Congress was Bernhard E. Fernow, who later became the first professional forester employed by the Federal government. Fernow had been trained at the Forest Academy of Munden in Prussia, and he discussed the development of forest care in Germany, pointing out that reforestation extended back in time as far as Charlemagne. The Forestry Congress also received a special communication from the Royal Chief Forester of the German Empire. The knowledge of Europe's experience encouraged American conservationists to seek similar government regulations and controls in the United States. But, at the same time, it was this aspect of the conservation movement that conflicted most with traditional American beliefs in the free exploitation of the national domain. Westerners' fears of Federal control seemed substantiated when in 1889 the American Forestry Association, successor to the Congress, and the American Association for the Advancement of Science urged Congress to set up forest reserves and temporarily withdraw all forest lands from sale. Finally, on March 2, 1891, in the Act to Repeal the Timber

Culture Laws, Congress authorized the President to establish forest reserves from the lands in the public domain. Both Presidents Harrison and Cleveland took the needed action, but Cleveland in 1893 refused to withdraw more forest land until adequate provision was made for the care of the existing government reserves. This situation was partly corrected by the Act of June 4, 1897, providing that "No public forest reservation shall be established, except to improve and protect the forests within the reservation, or for the purposes of securing favorable conditions of water flows, and to furnish a continuous supply of timber for the use and necessities of the citizens of the United States." [9]

By the turn of the century conservationists could be divided into two groups: those who favored a planned and scientific use of natural resources; and lovers of nature who wished to preserve the natural landscape unspoiled, as nearly as possible, by civilization. For example, nature lovers and big-game hunters wanted the national forests kept inviolate as parks or game preserves, while stockmen and lumber companies favored the use of these lands for grazing and commercial timber. Both groups of conservationists had already achieved some success in their attempts to influence government policy. Wildlife preserves and national parks helped keep nature in its pristine state. And, at the same time, conservation was becoming closely tied to the progressives' urgings of an efficient use of natural resources under government controls. Progressives and conservationists criticized the traditional American emphasis on competition as leading to overproduction and consumption, with resultant waste and inefficiency. Such practices were linked to monopoly, but monopolists were also being prosecuted for limiting production, although in this they would seem to have been supporting conservation. As one authority, in commenting on Theodore Roosevelt's inconsistent efforts, observed: "The trust-buster and conservationist are strange bedfellows." [10]

Although some students of the conservation movement have argued that the "organized conservationists were concerned more with economic justice and democracy in the handling of resources than with mere prevention of waste," [11] it is perhaps more correct to say that the conservationists advocated the kind of efficiency and scientific management which they believed could be best achieved through government regulation and control. Frequently hostile to laissez faire and traditional American individualism, conservationists accepted Theodore Roosevelt's concept of a regulated monopoly, and the conservationist gospel of efficiency became a part of both big government and big business. "The conservation movement," one of its recent students has declared, "did not involve a reaction against large-scale corporate business, but, in fact, shared its views in a mutual revulsion against unrestrained competition and undirected economic development. Both groups placed a premium on large-scale capital organization, technology, and industry-wide cooperation and planning to abolish the uncertainties and waste of competitive resource use." [12]

Considered in the above terms, conservation was an exercise in public and business administration, rather than an example of democracy at work. It was not a case of the people versus selfish interests. Frequently big business favored conservation, while small grass-roots farmers and entrepreneurs distrusted the degree of centralized control it involved. Finally, conservation became a matter of patriotism and of national security in the midst of growing imperialistic world rivalries. The great exponent of this side of conservation was Gifford Pinchot, who declared that "the conservation of natural resources is the basis, and the only permanent basis, of national success. . . . The planned and orderly development and conservation of our natural resources is the first duty of the United States." [13]

Gifford Pinchot, who has been called America's First Forester, was born a Connecticut Yankee on August 11, 1865. His parents were wealthy and on the father's side descended from French Huguenots who had settled in Milford, Pennsylvania. After studying botany at Yale, Gifford, who had determined at his father's suggestion upon a career in forestry, traveled in Europe where he could observe at first hand the latest scientific and practical methods in forest care. On his return to the United States, he inspected some of the timber lands of the Phelps Dodge Company, and then he became forester in charge of George W. Vanderbilt's Biltmore estate in North Carolina. On the Vanderbilt lands, he was given the task and opportunity "to prove what America did not yet understand, that trees could be cut and the forest preserved at one and the same time." [14] His success at Biltmore and in other consulting positions led to an appointment in 1896 to the National Forest Commission. This gave Pinchot a foothold within the formal conservation movement and valuable contacts in the Federal government, so that on May 11, 1898, he was named head of the Forestry Division of the Department of Agriculture.

In his new position Pinchot had to move carefully at first. Since forest reserves were under the jurisdiction of the General Lands Office of the Department of the Interior, he was literally a forester without forests. Finally, in 1905, after a long struggle in Congress in which he was aided by President Roosevelt, all forest reserves were placed under the jurisdiction of the Department of Agriculture and Pinchot's Bureau of Forests. In the meantime Pinchot had begun a campaign of education to persuade private owners that their timber lands could be scientifically cut to yield a continuing profit without destroying the forests. In behalf of his ideas he lectured widely and wrote for various national magazines. In 1899 he published the first part of *A Primer of Forestry*, of which eventually more than one million copies were circulated.[15]

A great help to Pinchot, and a tremendous encouragement to the cause of conservation in general, was Theodore Roosevelt's entrance into the White House. Unsparing in his criticism of the so-called nature fakers, the President was keenly interested in the preservation of the natural beauty as well the rich resources of the country. On his first presidential visit to the Far West in 1903, he toured North Dakota and the Yellowstone, accompanied by John Burroughs. In California, he and John Muir spent a weekend in the forests of the Yosemite, where the naturalist had a good chance to talk freely with the President concerning his ideas on conservation. Although Theodore Roosevelt lent a sympathetic ear to the nature lovers like Muir and Burroughs—setting aside the Grand Canyon as a national monument in 1908, for example—his major contribution to the conservation movement was the practical political support that he gave in Washington to leaders of the cause like Pinchot. The latter, in turn, although he deprecated the sentimentalism of those like Muir, was considerably aided in securing popular backing for conservation by the activities of the nature organizations, civic improvement groups, and garden clubs that were springing up across the nation.[16]

Like Roosevelt, but in advance of most public officials of his day, Pinchot saw clearly the close relationship between propaganda, control of public opinion, political lobbies, lawmaking, and appropriations. To mobilize public opinion behind the cause of conservation, he used the Bureau of Forests to conduct an extensive publicity service. The Bureau provided technical information, lantern slides, and other materials for schools and associations, and for lecturers, teachers, and writers, while Pinchot and his subordinates themselves did much writing and speaking and issued frequent news releases. In a letter in 1903 Pinchot pointed to the value of good public relations. "Nothing permanent," he said, "can be accomplished in this country unless it is backed by a sound public sentiment. The greater part of our work, therefore, has consisted in arousing a general interest in practical forestry throughout the country and in gradually changing public sentiment toward a more conservative treatment of forest lands." [17]

Pinchot was able to give practical backing to his philosophy of public administration by making the Forest Service into an extraordinary example of an efficient bureaucracy. He surrounded himself with capable men and devoted his own full time to the job. Wealthy and as yet unmarried, he gave his whole life to forestry. For some of his subordinates he provided additions to their government salaries, and he contributed four times his own salary to government work.[18] At the same time the Forestry budget increased majestically. In a period when there was a threefold growth in the number of forest preserves and in the acreage set aside by the President on Pinchot's recommendations, the rise of Forestry appropriations was even more spectacular: from $28,520 in 1899 to $3,572,922 in 1908. Over these same years, approximately 46 million acres of forest lands in 41 reserves trebled to 150 million acres in 159 national forests.[19]

With all his enthusiasm for forestry, Pinchot was sometimes accused of going too far. Congressional and newspaper critics argued that most of the Forestry money went for propaganda and not directly into conservation. Secretary of Agriculture James Wilson wrote Pinchot that Congress felt his publicity was an effort "to set up a 'forest fire' behind them." In 1908 Congress adopted as an amendment to the agriculture appropriations bill a provision that no part of the funds should "be paid or used for the purpose of paying for in whole or in part the preparation or publication of any newspaper or magazine article." Pinchot himself recognized the need for tact and political caution. To a field official he wrote: "In Government work the soft pedal is essential. 'Step softly and carry a big stick.' But don't use the stick. It is especially important that irregular methods, such as you used in several cases, and quarrels with local residents should be avoided." [20]

In most of its actions the Forest Service was upheld in the courts, but Pinchot, believing in a broad and elastic interpretation of the Constitution and of governmental powers, was never unaggressive. "It is the first duty of a public officer to obey the law," he declared. "But it is his second duty, and a close second, to do everything the law will let him do for the public good, and not merely what the law directs or compels him to do. Unless the public service is alive enough to serve the people with enthusiasm, there is very little to be said for it." Pinchot also gave his views of public service interesting application by reserving potentially valuable water-power sites under the guise of making them Forest Ranger stations.[21]

Conservation, of course, included more than scientific forestry practices. The preservation of water and mineral resources was also a matter of growing concern in the 1900's. Under Theodore Roosevelt the government began to withdraw coal lands still within the public domain from public sale. Of still greater interest to the West were water resources—both for irrigation and power. To cope with the problem of the vast arid areas of the Western United States, which had attracted the attention of Major Powell in the 1870's, Congress at first confined itself to donating land for irrigation purposes. The next step came with the Newlands Act of 1902, creating the Reclamation Service in charge of Frederick H. Newell, a government conservationist second only to Pinchot in his influence on the President. The Newlands or Reclamation Act assigned the receipts of land sales in the arid states to the construction of storage reservoirs or other permanent irrigation works. The Federal government by thus providing the land and water for irrigated farming encouraged anew the possibility of homesteading. In addition, the forestry program helped to build natural watersheds to conserve runoff waters and to prevent excessive erosion of the soil.[22]

Finally, to establish some over-all plan in regard to the water resources of the nation, President Roosevelt created the Inland Waterways Commission. Stating that his action was "influenced by broad considerations of national policy," the President pointed out that the time had come to merge "local projects and uses of inland waters in a comprehensive plan designed for the benefit of the entire country." As Roosevelt implied, the Commission was really intended as the beginning of an integrated conservation policy. Further indication of this broad national purpose for the Commission was the prominent part in its work assumed by the conservationist leaders Pinchot and W. J. McGee. Although the President credited Pinchot with the idea for the Commission, the latter indicated that it was McGee who was the real author. McGee, whom Pinchot called "the scientific brains" of the conservation movement, also provided it with a rationale or plan of action. He had served with Powell in the Geological Survey and, like Powell and Pinchot, was a strong advocate of positive government action. He believed that conservation should be considered as a whole comprehensive movement and not be limited to any single natural resource.[23]

To further a broad conservation policy and carry forward the idea behind the Inland Waterways Commission, Pinchot urged President Roosevelt to call all the state governors together in a national conservation conference. Roosevelt heartily approved this suggestion, and in the final year of his second administration the early conservation movement also reached its climax in the famous Conference of Governors at the White House in May, 1908. By this time, as Pinchot later recalled, conservation had come to have the connotations and unity later ascribed to the movement. It emphasized the comprehensive and well-planned management of all natural resources according to sound ethical and economic standards.[24]

Pinchot's utilitarian definition embraced the practical governmental aspects of conservation to which the agenda of the Conference of Governors was confined. With the help of McGee and Commissioner Newell, he dominated the planning for the Conference and served as its chairman. When Congress withheld an appropriation, Pinchot, according to Gilson

Gardner, a reporter friendly to the progressives, paid the Conference expenses and with his mother entertained one thousand guests at a reception in the Washington family mansion.[25]

The two major themes of the Conference were the impending depletion of natural resources and the necessity of their conservation as a matter of national patriotism. In the words of the President to the governors and their advisors and guests: "I have asked you to come together now because the enormous consumption of these resources, and the threat of imminent exhaustion of some of them, due to reckless and wasteful use, once more calls for common effort, common action." McGee, who likened the Conference to a Second Declaration of Independence, had prepared invitations for officers of a number of patriotic societies having nothing to do with conservation. Although the President, believing such invitations inappropriate, personally destroyed them, he nevertheless called conservation fundamentally a question of morality and patriotism. The American people, he feared, did not understand conservation as a "problem of national efficiency, the patriotic duty of insuring the safety and continuance of the Nation." [26]

Roosevelt's reference to a coming scarcity and to the citizen's patriotic duty to preserve national as well as natural resources provided themes for most of the other speeches at the three-day Conference. A few governors of Western states objected to the idea of a federally controlled conservation program despite the fact that support for this was one of the practical political objects of the Conference.[27] And Edmund J. James, president of the University of Illinois, raised the fundamental question of whether, in the concern over conservation, the American people were being persuaded to adopt unnecessarily restrictive government policies. With such policies in the past, Americans would not, he believed, have made some of their greatest advances in mining and agriculture. Moreover, James felt that the destruction of resources was being exaggerated while the possibilities of replacement were being ignored.[28] James' criticism was a minority view at the Conference of Governors, although Pinchot and others agreed with his contention that the intelligent, scientific use of resources was not the same as waste.

The Conference of Governors led to the establishment of a National Conservation Commission as a coordinating and fact-finding body. Also, in December, 1908, a Joint Conservation Conference was held between selected state and Federal officials. By this time conservation had become a well-organized movement. The older groups of naturalists with their nature societies and clubs were overshadowed by the new practical, political conservationists led by Pinchot and his fellow government officials. It was not accidental that many of these men were lawyers. After 1909 the rallying point for the conservationists was the National Conservation Association, which included a number of leading progressives in its membership and which became an effective lobbying body in Washington.[29]

Conservation now involved sizable political and business interests which ignored the older concept of conservation as a balance of nature. Conflicts like the famous Ballinger-Pinchot dispute brought competing interests over the use of resources into the open. But, even more significant was the way in which both government and business were coming to accept conservation in terms of scientific efficiency. Conservation to insure profits and national security was the new progressive goal. In the words of Pinchot's *Fight for Conservation*, "The central thing for which Conservation stands is to make this country the best possible place to live in, both for us and for our descendants. It stands against the waste of the natural resources which cannot be renewed, such as coal and iron: it stands for the perpetuation of the resources which can be renewed, such as food-producing soils and the forests; and most of all it stands for an equal opportunity for every American citizen to get his fair share of benefit from these resources, both now and hereafter." To this Pinchot added: "Conservation stands for the same kind of practical commonsense management of this country by the people that every business man stands for in handling of his own business." [30]

The early conservation movement which came to fruition in 1908 was a highly successful, practical propaganda effort intermixed with considerable idealism. It was able to persuade the American public that the natural resources of the West should belong to the nation and to the people as a whole, rather than to the states or to individuals and corporations. Conservation was thus in logical accord with the Roosevelt era, providing a good example of progressivism at work. As later difficulties testified, it also was part of the problem of reconciling an efficient bureaucracy and strong central government with individual and local interests. Finally, and perhaps most important, it illustrated the President's nationalistic approach and his concern with national security.

The 1900's witnessed America's increasing intervention in international affairs. Involved as never before in imperialist world competition, the American government for the first time gave serious consideration to the relationship of natural resources and national security. At the same time the needs of modern industrialism and filling-up of the West gave rise to fears of an impending scarcity of vital resources.

Conservation, as defined by Pinchot and his colleagues, had come to mean, not the effort to achieve a balance with nature, but the more efficient planned use of nature's resources. Interpreted in this way, it seemed to provide a popular scientific answer to the new national problems of the twentieth century. It appealed, not only to the progressive reformers' nationalism and patriotism, but also to their interest in social control and planning. It was democratic in a nationalistic rather than individualistic sense. In the words of Charles R. Van Hise, president of the University of Wisconsin, and author of the first history of conservation in the United States, "He who thinks not of himself primarily, but of his race, and of its future, is the new patriot." [31] This nationalism inherent in progressivism and conservation was to win even greater acceptance with the coming of the First World War, and later the New Deal. But the destructiveness of the war, at least, did little to conserve resources or bring man into closer harmony with his natural environment as distinct from his national government.

Conservation, considered only in these national terms, avoided the issue of civilized man's increasing helplessness before the interdependent structure of modern society. A war or bad harvest in one part of the globe now affected other areas. Conservation therefore was part of a larger problem that transcended national interests or rivalries. It also involved more than the question of private versus government use of natural resources, or their control by legislation. Future generations, as Professor Nathaniel S. Shaler pointed out in 1910, in his book *Man and the Earth*, would observe in retrospect how the current generation had used our and their environment. "They will date the end of barbarism from the time when the generations began to feel that they rightfully had no more than a life estate in this sphere, with no right to squander the inheritance of their kind." [32]

IX

A PLANNED SOCIETY

Conservationist ideology was one of the component parts of the idea of a planned society. As the old hope of achieving a natural automatic balance and harmony faded, new visions of reaching the same goal via conscious scientific planning emerged more brightly. The Utopian dreams of agrarian reformers and romantic transcendentalists seemed to hold less and less reality in the industrialized mechanized world of the twentieth century. But the old ideal of a proper balance between the forces of man and nature might nevertheless be attained through purposeful planning and effort. The conservation movement was a first step in this direction, providing a common ground for the growing faith in science and government regulation that characterized so much of the Progressive era. This degree of scientific planning and centralized control involved in conservation was often overlooked in the excitement of the nationalism and patriotism invoked by Roosevelt and Pinchot to gain popular backing for their cause. But the conservation movement was one of the bridges from the individualism of the nineteenth century to the collectivism of the twentieth.

Socialism was never very explicit or popular in America, and a planned society, or socialism by indirection, was often looked upon as a means rather than end in American thinking. Thus planning was accepted to save the forests, to win the war, or to combat the depression. Even in the midst of the concentrated planning carried on by the American government in the Second World War, the public still fancied itself in the image of an older individualist America. Probably this image was weakest in the 1930's, during the New Deal, when a wide variety of American opinion turned to the concept of a planned society, not only in desperation, but also with considerable enthusiasm. Important to New Deal planning was the idea of balancing production and consumption to preserve the price structure and conserve vital natural resources. This was a goal forged in the heat of the depression, but as emergency relief and recovery measures yielded the stage to long-range reform programs, some New Dealers also envisaged the more positive ideal of a cooperative planned society of self-sufficient homes and communities in which people would live in an almost agrarian harmony with their environment.

A century before the New Deal the economist Henry C. Carey and some of the Whig Party leaders embraced the idea of a harmony of economic resources and interests. Later, this concept was further developed as a part of the reaction of certain economic and social thinkers against the notion of an automatic natural evolution in the relationship of man and his environment. Harmony, it was believed, was a result of man's progress and technical ingenuity in utilizing nature's resources. Darwinian evolution in the light of new data and modern needs seemed to indicate the necessity of change by planned reform rather than by natural law.

One of the most important advocates of social planning in

terms of natural resources was Lester F. Ward. A pioneer sociologist who offered his collectivism too early to reach a properly receptive audience, Ward has nevertheless been cited as having contributed more than any other single individual to formulating the basic pattern of the American concept of the planned society.[1] Like Major Powell, Ward served in the Civil War and then entered government employment. This experience, as in the case of Powell, may have made him less suspicious and critical of government regulations and controls. Favored by his friend Powell with an easy work schedule as a paleontologist in the United States Geological Survey, Ward was able in 1883 to complete his first book under the significant title *Dynamic Sociology*. Sociology, Ward believed, was the analysis of a society actively changing; it was not a mere synthesis of the pure or static factors in society.

While studying the physical and social evolution of mankind, Ward decided that it was the mind which marked the great break between man and the animal kingdom. He denied the materialist contention that civilization was the outcome of crude physical needs, and instead developed the thesis illustrated by the title of his book *The Psychic Factors of Civilization.* Man's inventiveness and will power were more important than natural forces in changing society. The methods of nature were too slow. But man, rather than waiting for nature to effect modifications, was able to relieve environmental pressures by substituting his own free will and activity. Thus man changed his environment, radically and drastically, and in contrast to the animal kingdom was not, in turn, greatly affected by that environment. The efficacy of the free and active mind led Ward to the conclusion that purposeful planning through the collective agency of the government was necessary for continued future progress. An opponent of laissez faire, but dubious of socialism as an a priori unscientific philosophy, Ward espoused collectivism to the extent that it should prove needed, recognizing at the same time that complicated economic planning would put a strain on popular democratic government. "The individual has reigned long enough," he declared. "The day has come for society to take its affairs into its own hands and shape its own destinies." [2] Ward's assault on individualism and the pragmatic nature of his collectivism explain a large part of the appeal of his sociology to progressives of the Square Deal and the New Deal.

Successor in many ways to Ward in the general direction of his thinking was Simon N. Patten, professor of economics at the University of Pennsylvania before the First World War. Patten's ideas were an interesting example of the continuities of thought. Patten himself was impressed by the writings of Carey. Meanwhile, his own influence as a teacher rivaled that of his contemporary, William Graham Sumner, and "Patten men," especially Rexford Guy Tugwell, later became important in the councils of the New Deal. Despite the passage of time stretching from Carey through Patten, to Tugwell and the New Deal, there was a more than incidental connection between them. Patten, like Ward and Carey and later the New Dealers, was an economic optimist. Nature, he felt, was all right if man did not bungle it, but, with the growth of civilization, it was inevitable that the natural surplus of resources would diminish. Accordingly, it was necessary that man, artificially and by his own efforts, take steps to counteract the law of nature's diminishing returns and apply himself to the creation of a social surplus. "The situation, then," he wrote, "is this: the natural surplus is steadily decreasing, or at best it merely holds its own; the social surplus is the result of conscious effort, which must so arouse mental traits that natural decreasing returns become socially increasing returns." [3]

In America there were still abundant natural resources, and the problem was to translate these into a social surplus in order to insure a continued and stable prosperity. Hitherto the era of economic individualism had led to the exploitation of natural resources, to poverty, and to economic instability. But now, Patten continued, "The final victory of man's machinery over nature's materials is the next logical process in evolution, as nature's control of human society was the transition from anarchic and puny individualism to the group acting as a powerful, intelligent organism. Machinery, science, and intelligence moving on the face of the earth may well affect it as the elements do, upbuilding, obliterating, and creating; but they are man's forces and will be used to hasten his dominion over nature." [4]

Patten, like Ward, emphasized the psychic factors in progress and the need for positive planning. At the height of the progressive movement, just before the First World War, Patten's point of view was widespread in political and reform circles. In a classic exposition which attracted the attention of a number of the leading progressives, Herbert Croly, a founder of the *New Republic* magazine, argued that the promise of American life, based on its rich resources, was no longer to be fulfilled automatically but would have to be achieved more and more by disciplined economic planning and government regulation. Although Croly's popularity waned, his books, *The Promise of American Life* and *Progressive Democracy*, published in 1909 and 1914, gave the most detailed expression to ideas later incorporated in the New Deal.

In some phases of the conservation movement, and again as part of the process of economic mobilization in the First World War, the concept of balanced planning urged by Ward and Patten and Croly began to take hold. But it was the depression and New Deal that really popularized planning. After 1929 the practical situation encouraged the reformulation and reassertion of ideas deemed apposite. The multitude of voices calling for some sort of planning in the 1930's often bore little relation to older visions of an ideal harmony between man and nature. But underlying these new calls to action and reform was a belief that social and economic forces were sadly out of balance, and that some kind of an enforced adjustment was necessary. By the early 1930's the concept of a planned society was achieving widespread support. National groups, including the Chamber of Commerce, American Federation of Labor, and Federal Council of Churches, took official stands in favor of comprehensive economic planning, a position also backed by representative figures in business and industry such as Daniel Willard, Gerard Swope, and Owen D. Young.[5]

Indicative of the times and prevailing mood was the statement by Nicholas Murray Butler, president of Columbia University, in his address in 1931 before the American Club of Paris. Butler told his audience that the disastrous effects of poverty and unemployment were threatening the whole fabric of Western civilization. Believing that world statesmen could no longer delay the imposition of bold schemes for reform, he explained to his listeners that "if we wait too long somebody will come forward with a solution we may not like." Butler also ventured the opinion that "the characteristic feature of the experiment in Russia . . . is not that it is communist, but that it is being carried on with a plan in the face of a planless opposition. The man with a plan," he added, "however much we may dislike it, has a vast advantage over the group sauntering down the road of life complaining of the economic weather and wondering when the rain is going to stop." [6]

The depression popularized planning. Forgotten therefore, in the midst of the crisis, was the extent to which a measure of economic planning was already being practiced in business and government. Herbert Hoover, for example, although he later became a symbol of rugged individualism, was an early advocate of scientific planning. A highly successful mining engineer, Hoover during the First World War carried out the involved tasks of directing the activities of the Belgian Relief Organization and the United States Food Administration. As Secretary of Commerce under Presidents Harding and Coolidge, he developed an aggressive program to further American trade abroad. And as President he continued to encourage scientific research and fact-finding as a basis for policy-making. Opposed only to the notion of a highly centralized, government control of planning, Hoover stressed economic planning by way of the expansion of scientific knowledge and voluntary cooperative action. Thus Hoover came closest of the Republican leaders in the twenties to accepting the idea of social and economic planning. Indeed, his fondness for appointing research and study commissions was roundly criticized in the presidential election campaign of 1932, and the Democratic Party platform that year at least equaled the Republican in its adherence to traditional laissez-faire doctrines.

Though the candidate of the Democrats in 1932, Franklin D. Roosevelt was not in complete accord with the Party's platform or with many of its traditional, individualist tenets. One of his enthusiastic admirers later recalled that Roosevelt's first historic achievement was that, in his campaign for the presidency, he "raised economic and social planning to the status of a recognized national policy." [7] Even before his nomination Roosevelt in pleading for "a concert of action, based on a fair and just concert of interests," announced: "I am not speaking of an economic life completely planned and regimented. I am speaking of the necessity, however, in those imperative interferences with the economic life of the Nation that there be a real community of interest. . . . I plead not for class control but for a true concert of interests." In a major campaign address before the Commonwealth Club of San Francisco in September, 1932, Roosevelt dismissed the American tradition of limited government as outmoded. He criticized what he felt was the haphazard nature of American industrial and economic life in the past. The times now demanded that the government intervene to assist business in formulating "an economic constitutional order." [8]

Although the New Deal quickly became synonymous with the whole concept of planning, Roosevelt at times liked to stress his pragmatism and opportunism, and his lack of fixed devotion to any one set of ideas. But implicit in the New Deal program was the view that America was economically and socially a sick society, and that a comprehensive planned regimen of therapeutic reform was necessary. While eschewing the radical programs favored by the left- and right-wing governments of Europe, the Roosevelt administration was nevertheless convinced that voluntary private planning could not restore the country's economic balance. America, in the eyes of Roosevelt and his advisers, had achieved a mature economy in which the major problem was no longer one of increasing production, but of the proper distribution of consumer goods.

Concentration of wealth had resulted in overproduction, or at least in an inability of the masses of the people to consume their own factory and farm output. By economic and political reforms the New Deal hoped to redistribute the country's wealth and restore purchasing power among the lower classes. The United States, then, without sacrificing capitalism, would be able to achieve a balanced economy, and at the same time free itself from dependence on world export markets. Foreign trade implied exploitation and endangered both economic and political harmony. It invited the risk of war.

The interpretation of history and economic policy set forth under the New Deal was popularized in the writings of a number of sympathetic journalists and administration leaders. In many ways the most thoughtful of these documents, and the one of most lasting interest, was Secretary of Agriculture Wallace's volume *New Frontiers*. Wallace took an historical view of the American scene during the depression. American democracy, he feared, was faced with a crisis not unlike that of the Graeco-Roman world in the time of Augustus. The close of the nineteenth-century westward movement required the substitution of new frontiers if American and Western civilization were to survive. The American people would have to acquire through science and government the kind of harmony that had formerly been achieved through natural environmental factors. "An enduring democracy can be had only by promoting a balance among all our major producing groups, and in such a way as does not build up a small, inordinately wealthy class. . . . The complexities and the confusion of modern civilization are such that legislators quickly forget objectives of social and economic balance, and give way to the special pressures of the moment." [9]

Wallace acknowledged that there was "something wooden and inhuman about the government interfering in a definite precise way with the details of our private and business lives." But, he noted that the World War had given a tremendous impetus to comprehensive planning. Except in the matter of natural resources, however, he saw no reason why the United States should have to adopt a system of over-all planning, unless faced by an emergency.[10]

Wallace particularly stressed economic and social planning as a governmental device to thwart special vested interests and achieve a democratic balance. Denying that the N.R.A. and the A.A.A. were intended as permanent controls, he asserted: "We are committed to getting the farmer, the laborer, and the industrialist such share of the national income as will put each in a balanced relationship with the other. Without such balance the foundation of the state sags." It was Wallace's view that the regulatory functions of the economic market place had broken down in both national and international trade. Government, he argued, must provide the substitute, and its regulation would have to be extended into new areas. "If our civilization is to continue on the present complex basis," he wrote, "modern democracy must make rules of the game that go beyond tariffs, monetary policy, freight rate structures, taxation and similar policies which have long concerned the central government. The new rules must also get into fields more directly concerning harmonious relationships between prices, margins, profits and distribution of income." This, he felt, represented the essential challenge to Americans, whom he called "new frontiersmen." Wallace also pointed out that "in order to build the ideal democracy we need more people who know and are willing to pay the price that must be paid to bring about the harmonious relationship between this nation and other nations." [11]

Wallace believed that the New Deal both required and encouraged the subordination of the individual to the group interest. In the age-old fashion of American Utopians, he was confident that eventually this temporary loss of individuality would lead to a more stable and permanent individualism. Looking backward as well as forward, Wallace hoped that the joys of the older small-town or agrarian community could be recreated in the new planned frontier of the future. "The keynote of the new frontier is cooperation just as that of the old frontier was individualistic competition. . . . Power and wealth were worshiped in the old days. Beauty and justice and joy of spirit must be worshiped in the new." With economic need eliminated in the new communities, much of the traditional pettiness of small towns would disappear. Wallace pointed out that it might take a greater degree of centralized power along the lines of the New Deal experiments to achieve the eventual goal of a greater decentralization of American society. But, accepting this paradox, he concluded: "I cannot help feeling that eventually the physical manifestation of the new frontier will consist in considerable measure of thousands of self-subsistence homestead communities properly related to decentralized industry." [12]

Wallace's hope that the collectivist methods of the New Deal might in time recreate a nation of decentralized agrarian communities was reminiscent of the Jeffersonian dream. This notion was also an illustration of some of the confusion and contradiction in the unfolding New Deal program of legislation. The specific measures enacted, especially during Roosevelt's first administration, were unprecedented both in volume and significance. Never before, it seemed safe to say, had legislation cut so deeply into the economic life of the nation, even though the basic capitalistic structure of the country was not overthrown. Paradoxically, vast projects verging upon socialism, like the T.V.A. and the Social Security program, escaped major criticism, while smaller efforts to encourage rural resettlement and subsistence homesteads were bitterly attacked as Utopian Socialist fancies. Yet, in some ways, the latter were attempts to restore the old agrarian society and came closest of all New Deal experiments to reviving the notion of an ideal balance between man and nature.

The economic depression, with its mounting volume of industrial unemployment, encouraged a movement back to the land or to what Ralph Borsodi, one of the pioneer decentralists and critics of technocracy, called "Flight from the City." Although Americans had overrun the best lands in the West, at the same time abandoning hundreds of thousands of marginal acres in the older states of the East, this back-to-the-land movement was continually nourished for over a century by the Utopian schemes of a variety of reformers, agrarians, individualists, and socialists. In New York State, where the rate of abandonment of farm lands had averaged 100,000 acres a year since 1880, Governor Franklin Roosevelt was attracted to the idea of using these tax-delinquent lands for reforestation or subsistence homesteads. In the Federal government, Elwood Mead, Commissioner of the Bureau of Reclamation from 1924 to 1936, had long advocated irrigation and reclamation of the soil as a means of rebuilding a healthy rural society and as a way of putting families back on the land.[13]

Under the New Deal the agrarian visions of Mead and others were united with a comprehensive program of government planning for the natural resources and economy of the country. Conservation and reclamation of the land were now tied to the idea of rural resettlement and subsistence homes. But instead of the widely scattered, isolated, and worn-out farms of the past, the new rural communities were to be patterned on the style of a European village. With the mobility afforded by the gasoline engine, the farmer could live in town and still work his outlying acres. In a rural setting it might become possible to enjoy the sense of economic and social balance and security made possible by part-time farming combined with industrial employment.

Though the details of all these ideas had yet to be worked out, official encouragement was supplied by a little-noticed section in the National Industrial Recovery Act of May, 1933, which provided an appropriation of $25 million to aid individuals in the "purchase of subsistence homesteads." In spending this appropriation a conflict over the costs of the houses quickly arose between those who saw them as a simple relief measure and those who envisaged the program as a demonstration of a new way of life. After the demise of the N.R.A. the subsistence homesteads program was transferred to the newly created Resettlement Administration, with only $8 million of the original $25 million spent. Under the direction of Rexford Tugwell the Resettlement Administration became one of the most controversial of all the New Deal agencies. Its functions, as detailed in the President's executive order, embraced the "resettlement of destitute or low-income families from rural and urban areas, including the establishment, maintenance, and operation, in such connection, of communities in rural and suburban areas." Tugwell was in sympathy with the agrarian ideals of the subsistence homesteads program, but he was also among the most avowedly radical collectivists of all those high in the councils of the New Deal. Bitterly criticized by local interests and conservative farm organizations, the rural resettlement and subsistence homestead programs were never able to realize the high expectations of their sponsors, although ultimately over one hundred various types of projects were built.[14]

According to a recent historian of the program, its initiation had been influenced by a quasi-Jeffersonian agrarianism, but its development reflected an open break with traditional individualism and pointed clearly toward collectivism. Although one of the smaller New Deal ventures in planning, it was a significant experiment and a focal point of ideological clash and criticism. To a dedicated planner like Tugwell, who was at the center of the fight over the Resettlement Administration, it was the collectivist community aspects of the program that made it worthwhile. To agrarian individualism and the harmony of man with nature were to be added the planned social services and tightly knit gregariousness of the village community. Moreover, in the eyes of many of the conservationists in the New Deal, the subsistence homestead program was merely one of many coordinate problems, including soil erosion, submarginal lands, and ignorant poverty-stricken farmers. In 1937, therefore, the Resettlement Administration was merged into the Farm Security Administration of the Agriculture Department, and the rural subsistence homestead program gradually disintegrated.[15]

In the period of the New Deal the increasing complexity in the relations of man with nature and the growing difficulty of achieving any sort of balance and harmony were further illustrated by the elements of confusion and paradox in the development of the Tennessee Valley Authority. The original cooperative collectivist goals of the New Deal proponents of the T.V.A. had to be shelved in response to political attacks and local community pressures. Instead of becoming a center of comprehensive regional planning on a vast scale, T.V.A. became a means of cheap electrical power and improved river navigation which attracted industry to the area. In this way private enterprise was encouraged, and a different kind of economic balance was ultimately gained. According to Commissioner David Lilienthal, T.V.A. showed the possibilities of collectivism without the sacrifice of democracy. But to an ardent early New Dealer like Tugwell, who desired to carry the original T.V.A. collectivist planning into much broader areas, the compromises with local political sentiment were viewed as "an example of democracy in retreat." [16]

In the case of T.V.A. as well as other New Deal plans, war intervened to effect a drastic change of purpose. Instead of a democratic harmonizing of conflicting interests, and a better balance between man and the resources of nature, the war imposed a ruthless demand for goods and manpower. Quiet agrarian ideals, whether individualist or collectivist, could not survive total war. Governmental interest in conservation became, of necessity, an effort to harness the nation's resources more efficiently in terms of the war effort. In the midst of the military struggle there could be little or no concern with balance and harmony, or with the needs of future generations. The ideal planned society became the pragmatic nation at war. Thus T.V.A.'s vast facilities at Oak Ridge were used to manufacture atomic bombs—an ironic achievement "of turning to destruction the energies of a region that had been entered for the express purpose of conservation and development." [17]

Full United States participation in the Second World War brought about significant changes in New Deal thought. Rather paradoxically, as the economy became more and more collectivized to meet the challenge of totalitarianism and war, a certain disillusionment with over-all planning began to be evident. Although the depression had killed the old confidence in the automatic beneficence of an acquisitive society, the war experience did not arouse enthusiasm for centralized economic planning. If free exploitation of resources could no longer be accepted with equanimity, neither was there any great support for a completely ordered and regulated environment. The postwar years therefore were to be marked by a new spirit of humility in regard to the possibility of man controlling nature. Even the tremendous achievement of atomic energy, because of its enormous potentiality for universal destruction, hardly inspired optimism. Rather it indicated all the more the need of harmony in the relationship of man and nature, and the necessity of some kind of a balance between human and natural forces.

X

RESOURCES AND ENERGY

Resources and energy are man-made. They are the environment applied to the service of civilization. Unused by man, the riches of the natural world could hardly be thought of in terms of supplying resources and energy. Yet civilized man with all his constructive work was also a great destroyer. Especially in the twentieth century his technological achievements raised the possibility that the world's stock of resources and energy would eventually be exhausted. Even ever-greater industrial productivity and the fabulous energy released by nuclear fission did not allay such fears. Henry Adams' historic pessimism accordingly had its modern counterpart, even though Adams' notion that "the entire universe, in every variety of active energy, organic and inorganic, human or divine, is to be treated as a clock work that is running down," no longer received serious attention.[1]

Political anxiety over the resources-needs of the nation was evidenced in both the early conservation movement and the economic measures of the New Deal. Indeed, much of the long-range planning of the New Deal was directed to the attainment of a comprehensive resources program, balancing both human needs and natural wealth. New Deal visions of a planned society rested in large part on a multi-purpose type of conservation in which the economic requirements of the nation were to be carefully worked out in terms of the future.

The original conservation movement of the early 1900's had resulted in the Federal government gaining control of a large part of the remaining natural resources of the West. Thus forest and mineral lands were withdrawn from public sale and made permanent parts of the national domain. But except in the case of national parks and wildlife sanctuaries, there was no thought that the resources in reserved lands would not be put to eventual use. The major problem was to decide the terms and auspices under which these valuable natural resources could be made available.

Although the conservation movement had seemingly settled the question of public ownership, there was considerable concern among organized conservationist leaders over the exploitative effects of the First World War. Harry Slattery, executive secretary of the National Conservation Association, distrusted the industrialists who came to Washington as dollar-a-year men to help Wilson win the war, and Progressive Republicans were skeptical of the devotion of the Wilson Administration to conservationist ideology. Slattery and Pinchot, for example, expressed the fear that Secretary of the Interior Lane was "going to give away every thing in sight" before the war was over.[2]

Despite the fact that the United States and Germany were at war, conservationists pointed to Germany's careful use of its land and resources as a model for the United States. Richard T. Ely, the well-known professor at the University of Wisconsin, who had received his graduate training in economics in Germany, expressed the view that conservation was aided by the German idea of the national state "as an institution representing all the people, past, present and future, coupled with the idea of devotion to it as a sacred duty of the citizen." A laissez-faire philosophy was antagonistic to conservation which, Ely declared, "necessarily means more public ownership, more public business; this means a demand for better government; and this means giving men a real career in the public service." [3]

Conserving resources for possible military needs was, of course, a wartime fact of life. At the same time, however, war could never result in any real saving. Instead it put a premium on the immediate exploitation of resources, whatever the dangers of possible future scarcities. In the United States the magnified wartime needs of the government revived Western states' opposition to Federal control and ownership of the public domain. The Federal government accordingly experimented with a wartime system of leasing public lands and resources, but since the Western states did not share in the royalties, conservation from their standpoint meant that the section most concerned received neither the use of its lands nor any direct portion of the rentals or revenues they provided. However, government forest lands were already being lumbered on a selective basis, and in 1920 the passage of the Water Power Act and Mineral Leasing Act made possible the private use of natural resources under continued Federal ownership, and with special benefits accorded to the West.[4]

During the 1920's conservation was encouraged by postwar fears of an exhaustion of key mineral resources and also by growing demands for greater efficiency in government and business. The most immediate concern was over the depletion of the oil reserves of the nation. The tremendous rate of growth in the use of petroleum products was at first not matched by the discovery of new oil fields, and the government therefore withdrew from sale public lands where oil was known to occur. Provision was then made for the subsequent exploitation of the fields by leases to private producers. Under this kind of arrangement it was hoped that a better balance between production and consumption could be achieved. Some authorities also advocated conservation of oil and gasoline products by allowing a possible future scarcity to dictate price rises. Concern over the exhaustion of petroleum resources was not allayed when the Teapot Dome scandal revealed that important naval oil reserves had been improperly and wastefully leased to private corporations. Gradually, however, optimism over potential oil resources in the United States was restored with reports of the discovery of rich new fields both at home and abroad. By the 1930's, when estimates indicated that the petroleum supply would last for years, it was recognized that earlier thinking had been too pessimistic.[5]

The Teapot Dome scandal probably helped create a number of false impressions with regard to the government's role in conservation during the twenties. Although the earlier Theodore Roosevelt era continued to be publicized, there was no real subsequent reversal in the government's policy of controlling natural resources. In fact in some ways Federal authority was increased. In the areas of reclamation and water power, the government expanded its controls and became more than ever a competitor of private industry, even though the wartime Muscle Shoals Dam was not put into use again until the Franklin Roosevelt administration. During the 1920's there was also evidence that the need for conservation had come to be taken for granted by the public. Although President Hoover suggested turning over the remaining public lands to the states, the Department of Commerce under his secretaryship had championed the idea of conservation. Increasingly in the twentieth century conservation was looked upon as a matter of economy and planning synonymous with business and engineering efficiency. One world authority declared in 1933: "Today the conservation movement is led by sober business men and is based on the cold calculations of the engineers. Conservation, no longer viewed as a political issue, has become a business proposition." [6]

Under the New Deal the state of the country's natural resources was the subject of considerable important legislation. At the outset of the Roosevelt administration the establishment of the Tennessee Valley Authority represented an effort at broad multi-purpose conservation for an entire submarginal region. Thus it went far beyond previous conservation programs. Basically T.V.A. was created to provide cheap electricity for the area, but it also helped control floods and resulting soil erosion and made possible the rebuilding of watersheds and forest lands. Other similar hydroelectric projects were envisaged. The Hoover Dam on the Colorado River was completed, and on the Columbia River in the Northwest the Bonneville and Grand Coulee Dams were started. In 1934 the entire remaining 165 million acres of the public domain were withdrawn from possible sale to be developed as national forests and as grazing districts. Here the administration adopted the point of view that much of the grasslands region of the Great Plains, subject to dust storms and dry erosion, was unsuited to homesteading. At the same time, to try and alleviate these drought conditions, President Roosevelt in an executive order set aside funds to create a shelter belt of trees to break the prevailing winds and collect moisture.[7]

The shelter-belt idea, though never completely achieved, illustrated the administration's emphasis on soil conservation. Relief as well as reclamation provided an excuse for a number of projects, and the Civilian Conservation Corps was able to provide a necessary labor force. At first soil erosion was regarded as purely a conservation or reclamation project, but later the Soil Conservation Service became part of a comprehensive program of crop control and acreage restriction. Cutting down farm acreage, while increasing productivity and also encouraging a back-to-the-land movement, illustrated some of the cross-purposes of the New Deal agencies entrusted with the care of the nation's resources. This situation led to a demand for an interrelated conservation program that might help avoid some of the interagency budget struggles and jurisdictional problems between Agriculture and Interior, Army Engineers and Reclamation Bureau, or Soil Conservation Service and T.V.A. Henry Wallace, for example, in 1939 declared that instead of separate problems in forestry, wildlife, grazing, soil, and crop adjustment, "there is one unified land use problem. . . . This problem involves the whole pattern of soil, climate, topography, and social institutions; it has to do with social and economic conditions, as well as with the physical problems of crop, livestock, and timber production, and of soil and water conservation." [8] In theory a measure of coordination might have been achieved through such a body as the National Planning Board, later changed into the National Resources Board or Committee. But the Board's functions were largely advisory and technical, and it expired during the war.[9]

As in the case of the First World War, the Second World War encouraged over-all planning and the stockpiling of critical natural resources. But the conservation of resources, despite theoretical government concern, became a practical impossibility. Indeed, in view of the extravagant realities of modern total war, governmental criticism of private waste and attempts to enforce saving by legislation and administrative control verged upon the insincere or dishonest. There was no escaping the blunt fact that governments everywhere, in their roles as war machines, were the greatest single exploiters of the natural world. Compared to their organized destruction, the selfish misuse of nature by private individuals or corporations was a minor problem. But, at least, the Second World War was not fought without exercising a certain sobering effect upon public thought. As never before in American history the postwar generation gave a sympathetic hearing to the scientists and economists who discussed conservation in terms of its wider, more philosophical aspects. Going back to Thoreau and Marsh, these scholars raised again the question of the balance and harmony of nature and of how long man could live by exploiting, rather than adjusting to, his environment. The extent of the problem was posed by two authors connected with the Soil Conservation Service of the Department of Agriculture. After noting that "Man's habit of destroying the natural resources from which he lives is as old as civilization," they pointed out that "Conservation is not something that can be controlled exclusively by legislation. It is largely a way of thinking and a way of living. It is as fundamental as honesty and thrift." [10]

Topsoil and Civilization, the work from which this quotation is taken, was a good example of the postwar attention to the importance and complexity of the problem of man in relationship to nature. In the Preface to their book, authors Dale and Carter asserted that "With the progress of civilization, man has learned many skills, but only rarely has he learned to preserve his source of food. Paradoxically, the very achievements of civilized man have been the most important factors in the downfall of civilization." Historians, they pointed out, had largely ignored land use and its importance to civilization. "While recognizing the influence of environment on history, they fail to note that man usually changed or despoiled his environment." Primitive man did not upset the natural process of soil, plant, and animal growth, and like other animals he was forced to adapt himself to his environment to survive. With the advent of civilized man the thousands of years of the earth's soil-building process was reversed. The soil declined in quantity in most areas, with the decline speeded as the inventive genius of man devised new tools and techniques to hasten the process of soil depletion. Thus man's "intelligence and versatility made it possible for him to do something no other animal had ever been able to do—greatly alter his environment and still survive and multiply." [11]

Dale and Carter saw a more than casual relation between topsoil and civilization from the fact that the world's progressive civilizations had hitherto not been able to continue indefinitely in one locality. An important exception that they noted was the Nile River Valley, which continued to furnish a stable home for civilized man after some six thousand years, or until recent times, when man finally became civilized enough to upset the natural balance of the area. In the twentieth century the Nile River dams, built to prevent flooding and accomplish irrigation, also prevented the silting of the river basin below, thus depriving the Nile Valley of necessary minerals and humus and forcing Egyptian farmers to face lower yields for their crops. At the same time the headwaters of the Nile began to be disturbed with timber cutting and land erosion so that artificial lakes above the dams were filled with silt that had formerly been permitted to flow down into the valley. The situation, however, had reached such a point that without the dams the resultant sudden silting and floods would probably have destroyed the remaining agricultural value of the lower Nile.[12]

The Nile River Valley was still productive, although modern Egypt was a country facing grave economic problems. Other areas, which in the twentieth century were regarded as backward and underdeveloped, had once been the scenes of flourishing populations that had contributed more over a long period to civilization than the United States. America as a civilization was still an infant, and it was also true that it was using up its natural resources at a prodigious rate, far in excess of the usage of other parts of the modern civilized world, and much more rapidly than the ancient and classical civilizations. At least in classical antiquity technology had not reached the level where the process of environmental change was greatly accelerated. The fertility potential of the soil was thus rarely endangered; the plows and other agricultural tools were not strong enough to cause serious soil destruction.[13]

In marked contrast with the pattern of man's use of the land in antiquity was the example of American agriculture. The native Indian crops, which had been grown before Columbus in limited quantities, were expanded according to European standards of large-scale farming. The plowed fields producing corn and tobacco became badly depleted so that "Soil erosion in the eastern United States," one authority asserted, "has been more destructive in the past three centuries than it has been in northern central Europe since the dawn of the Christian era."[14] Another commentator expressed the view that corn, which was traditionally the most ubiquitous American crop, grown almost everywhere in the United States, had damaged the earth's surface to such an extent that it deserved to be regarded as more harmful to man than syphilis.[15]

As land passed out of cultivation in the East, it was brought under production in the Western states via irrigation. Most of the water for this irrigation had to be impounded in reservoirs. Building dams for irrigation and navigation was popular in the United States, especially during the New Deal, but it was by no means sure that the country was escaping the problems that were developing in connection with the similar efforts of the Egyptian government in the Nile River Valley. In 1937 the National Research Council observed, on the basis of government sedimentation surveys, that "83 per cent of all existing reservoirs in the nation are threatened with silt extinction within less than 200 years." The Council estimated that 38 per cent of all reservoirs had a life expectancy of only 1 to 50 years; 24 per cent, 50 to 100 years; 21 per cent, 100 to 200 years; and only 17 per cent more than 200 years.[16]

The erection of large dams, though subject to little public criticism on grounds of unwise conservation, was challenged in some scientific circles. At Princeton, New Jersey, the scholarly symposium devoted to the general question of "Man's Role in Changing the Face of the Earth" included the promiscuous damming of rivers as among the dangerous interferences with the balance of nature. In contrast, port and harbor works, "being protective and local in effect," were praised as "almost universally beneficial." George Perkins Marsh had reached this same conclusion much earlier, commenting that by such coastal improvements "man has achieved some of his most remarkable and most honorable conquests over nature."[17]

Experts of the United States Geological Survey concluded

on the basis of elaborate statistical studies that there was a limit to the practical gains which could be accomplished by building reservoirs on streams, and that some drainage basins in the West were already approaching that limit. Though the trend in dam construction was still upward, "the point of ultimate development for hydroelectric power, irrigation, flood control, and navigation may be seen on the horizon," and it was expected therefore that attention would shift to maintaining water supply and improving pollution control. Water storage reservoirs, of course, were aimed at achieving a relatively even flow of water, storing water from wet periods, and releasing it in dry ones. But reservoir capacities could also reach a point of diminishing returns in which losses by evaporation canceled any increase in over-all capacity. This, for example, appeared to be the case in the Colorado River basin, where additions to the existing capacity of 29 million acre-feet would be largely offset by a corresponding increase in evaporation.[18]

In American history the value of man's efforts to change his environment was best illustrated by reference to the Great Plains area between the Mississippi River Valley and the Rocky Mountains. In this region, within the last century, the buffalo herds had been exterminated and replaced by cattle ranching and wheat farming. The area also continued to be devastated periodically by drought and dust storms. Various solutions were offered—including reversion of the wheat fields to grasslands, or the possible irrigation of parts of the territory through the damming up of the water. Cattlemen naturally preferred to keep the plains as grasslands for pasturing their herds, while farmers desired enough water to enable them to grow tillable crops. From a scientific point of view, however, a case could be made for letting the lands revert to their natural state under the Indians. Contributors to the Princeton symposium, for example, pointed out that the extent of the earth's surface supporting natural vegetation had not changed much over the centuries although the composition had been substantially altered in historic time. Plants had always been the primary source of food for civilized man, and it was not likely that science and technology would free man from his dependence on vegetation. Most of the major crops needed by man had been domesticated in prehistoric times, but methods of use and cultivation had changed. The force of these generalizations could be applied to the Great Plains area of the American West. There the nomadic grazing of the buffalo did not damage the Indian habitat. The pastoralism for commercial purposes of the American sheep and cattle industry effected much greater changes, but it was probable that "Man's really significant alteration of the mid-latitude grasslands has occurred where he has destroyed and replaced them by plowing and planting." [19]

In opposition to this view that the Great Plains grasslands should be allowed to revert to something like their natural state, a historical scholar pointed out that long before modern man the area was probably swept by fire, drought, and dust storms. The winds blowing over these dried-out plains deposited their load of soil to the east in the wet areas of the lower Missouri and central Mississippi River Valleys. The rich agricultural lands of the states of Missouri, Iowa, and Illinois therefore perhaps owed their fertility in considerable part to the operations of the so-called dust bowl to the west. Hasty crisis-minded solutions, however appealing to farmers in the dry areas of the Great Plains, might not represent long-range conservation or the achievement of a sensible balance between man and his use of nature.[20] From this point of view proposals to dam the Missouri River portended an agricultural disaster by interfering with the moisture that the river's overflow normally brought to the soils deposited by the western winds. In the words of one critic, "We lose sight of the biological aspects of the situation because the technological aspects of building big dams look so wonderful, but man's engineering ambition to push the world around with a bulldozer is seriously disturbing when this biological performance is completely upset."[21]

Next to plowing man worked his greatest change upon the earth's surface through his controlled use of water. Against the usual interpretation of civilizations as rural or urban, a classification was also made between the hydraulic and nonhydraulic civilizations. Hydraulic civilizations which depended on irrigation experienced a more intensive type of agriculture, with much use of human labor but little technology. China was an illustration of such a civilization. By utilizing various methods of irrigation the Chinese had developed a type of agriculture that made European techniques seem primitive by contrast. Since the hydraulic civilization of China had maintained itself through several millennia, it was obviously "a going concern." "Yet in terms of human affairs, it was as costly as it was tenacious," and condemned countless generations of human labor to extreme drudgery.[22]

Modern civilization put its water supply to more diverse use than the Chinese. Although scientists discounted spectacular notions of widespread general climatic changes from water storage or drainage, there was much expert agreement over the importance of the quantity of ground water stored beneath the earth's surface. Destruction of the natural cover of the soil interfered with this storage of ground water and was dangerously lowering the water table in many parts of the United States. Restoration depended both on conservation of the soil cover and on a system of irrigation that used ground water stored in the months of peak rainfall.[23]

The chance of any really significant water conservation was almost precluded by the tremendous use and waste of water in the modern industrial urban civilization. In destroying the natural cover of the soil the farmer and lumbermen were joined by the city dweller. Paved streets and highways reduced the normal seepage of water into the soil and increased the runoff that flowed into the sewage system. The disposal of man's personal and industrial wastes, by contaminating his rivers and lakes, further limited the supply of usable water. In part the problem was alleviated by purification and reuse of some of the water, but many American communities were not willing to pay the high costs for treating sewage and industrial waste. Probably an eventual solution to man's constantly growing need of water would depend on future technological achievements in reducing the expense of the process of desalting sea water. Another less attractive possibility was a forced economy in water use as a result of the higher costs of the product. This would reduce conservation to a matter of economics and transportation, in which it was predicted that "Water will be used in those places and for those purposes which can best afford to bear the cost under prevailing conditions." [24]

Despite the problem of water and the other factors which, since the close of the frontier, had added to the farmer's costs and ended the days of his easy exploitation of the West, American agriculture continued to be highly productive. American farm output was the greatest in the world per *man*, but not per acre. Moreover, good land was scarce. Only one fourth of the 2 billion acres in the United States was arable, although this, in turn, represented a high proportion of the world's agricultural area. So far the ingenuity of the individual American farmer and the technological achievements of large-scale mechanical agriculture had overcome obstacles of the natural environment. How long this would continue was uncertain, but the comment that, given the discoveries of agricultural chemistry, soil was needed only to hold up the plants, though an exaggeration, gave a hint of the possible future.

The success of technology as applied to agriculture illustrated the folly of assigning definite limits to the contriving human brain. Short-run gains, however, might not represent long-term wisdom. Man was still dependent for his food largely on plant life grown in the soil. Overproduction on American farms as a result of agricultural chemistry, mechanization, and government subsidies provided surpluses that were in marked contrast with the scarcities of most of the rest of the world. History gave no assurance that the United States would not in time face the same problems of maintaining agricultural productivity that had beset other civilizations.

Even if farmers by a scientific and technological miracle proved able to feed the world's growing population, there still remained on the horizon the grave danger of the exhaustion of nonrenewable mineral resources. Again and again man's fertile imagination had devised substitutes for scarce natural resources, but the competition between intelligence and nature seemed almost daily to grow more keen and ruthless. Believers in the necessity of conservation were often concerned therefore over the rate at which the world was exploiting and destroying precious minerals and metals—substances which could not be grown or replaced except by the slow natural rebuilding of centuries.

The scholarly scientists gathered together at the 1955 Princeton symposium on "Man's Role in Changing the Face of the Earth" devoted especial attention to the problem of mineral resources, although participants differed as to how imminent was the danger of their depletion. Also, in 1952, a small group of interested conservationists set up an organization called Resources for the Future. With the help of a foundation grant this informal agency was able to arrange for a series of national resources conferences. Despite their mounting concern over the problem, scientists remembered that previous estimates of the exhaustion of coal and oil had, after all, proved faulty as new sources were located. Although coal and oil were probably adequate for the twentieth century, with the hope that chemistry and solar or nuclear energy would then take over, there was still no way of gauging accurately future consumption. For example, Sir Charles G. Darwin, the distinguished English physicist, in questioning even such limited optimism, voiced the opinion that growing organization and complication were increasing the entropy of the world. Half the coal consumed in the history of the world was burned by the United States in the last thirty years. "In this sense, then," Darwin concluded, "the United States in the last thirty years has done as much to increase the entropy of the world as the whole of the human race in the whole of the past. This is not a boast!" he added.[25]

Samuel H. Ordway, another of the participants at the Princeton meetings, and an official of the Conservation Foundation, on the basis of the relevant findings of the President's Materials Policy Commission, pointed out that "*The quantity of most metals and mineral fuels used in the United States since the first World War exceeds the total used throughout the entire world in all of history preceding 1914.*" What was needed, according to Ordway, was greater public support of conservation, not merely for scientific or economic reasons, but as an attitude or way of life. "With such an ethic there would be fewer unexpected children and fewer unneeded luxuries; gadgets would be made to last longer; there would be less waste. There would be increased productivity, less erosion and destruction of soil, less escape of valuable water, better forestry, more wildlife habitat, wise husbandry of all resources. The bases of prosperity could be preserved." The dreams of a Golden Age of plenty and leisure, Ordway felt, might better be transformed into the effort to achieve a Golden Age of conservation. "Then we can be done with this ridiculous insistence upon industrial expansion and with all unnecessary production. Our goal will become industrial stability. Our civilization, so-called, will have matured." [26]

The report of the President's Materials Policy Commission, published in 1952, gave statistics elaborating the tremendous United States consumption of materials—far greater than the rest of the world and at a rate threatening exhaustion. "The United States appetite for materials is Gargantuan—and so far, insatiable." For example, in 1950, as compared with 1900, the United States was taking six times more minerals, including fuels, from the earth. Government concern over the problem was illustrated by its increased expenditures for research dealing with vital natural resources. But this involved a paradox since the Department of Defense and the Atomic Energy Commission together accounted for almost 90 per cent of the government's research spending. And it was these agencies which were also the most prodigal users of precious raw materials. In 1942 industry was responsible for 64 per cent of all research, government 26 per cent, and the universities 10 per cent, with payment and performance roughly parallel. Ten years later the comparable figures were industry 41 per cent, government 56 per cent, and universities 3 per cent.[27]

In view of its own high rate of consumption, the sincerity of the government's interest in conservation could easily be questioned, but hardly any more so than the careless attitude exhibited in all the reaches of modern industrial civilization. A recent discussion, prepared for leaders of American industry, and devoted to studying the prospects in the next hundred years for man's natural and technological resources, emphasized the tremendous amounts of raw materials required to support a single individual in a highly industrialized society such as the United States. Each year each person in the United States consumed an estimated 1,300 lbs. of steel, 23 lbs. of copper, 16 lbs. of lead, 3½ tons of stone, gravel and sand, 500 lbs. of cement, 400 lbs. of clay, and 200 lbs. of salt. "Altogether over 20 tons of raw materials must be dug from the earth and processed each year in order to support a single individual in our society. And these amounts are steadily increasing." [28]

Harrison Brown, an outstanding American scientist and senior author of this survey, maintained on the whole a position of qualified optimism. Technology, in the face of the ever-growing rate of the world's population increase, he asserted, could only alleviate a situation in which "More than one half of the people of the world are hungry today." Overpopulation and war were the chief threats to a balance of civilization and nature. Since, if the latter disaster struck, it was very doubtful that the world could rebuild its complicated industrial order, war had to be avoided at all costs lest the world relapse into an agrarian society. Brown predicted that modern machine civilization would spread rapidly over the globe and become stabilized, or it would stand as a temporary Golden Age before a violent reversion to agrarian life. Brown's forecast posed the ironic possibility that through war or some other catastrophe man would halt his denudation of the environment and achieve at tremendous sacrifice and pain an enforced balance with nature. The alternative was the peaceful attainment of some kind of harmony or stability through voluntary control of population.[29]

Brown's general point of view was supported by other influential scientists. In 1956 a committee on the Social Aspects of Science of the American Association for the Advancement of Science stressed the imbalances being caused by the admittedly great progress of science. Citing especially the dangers stemming from nuclear radiation, chemical food additives, and failure to conserve natural resources, the report in concluding its recommendations for more research and a greater public understanding of science stated: "The growth of science and the great enhancement of the degree of control which we now exert over nature has given rise to new social practices, of great scope and influence, which make use of new scientific knowledge. While this advance of science has greatly improved the condition of human life, it has also generated new hazards of unprecedented magnitude. These include: the dangers to life from widely disseminated radiation, the burden of man-made chemicals, fumes and smogs of unknown biological effect which we now absorb, large-scale deterioration of our natural resources and the potential of totally destructive war. The determination that scientific knowledge is to be used for human good, or for purposes of destruction is in the control of social agencies. For such decisions, these agencies and ultimately the people themselves, need to be aware of the facts and the probable consequences of action. Here scientists can play a decisive role: they can bring the facts and their estimates of the results of proposed actions before the people." [30]

In terms of its potential for good or ill, no discovery had greater implications in regard to man and nature than the release and control of nuclear energy. An all-out nuclear war would probably destroy most biologic life and the essentials of civilization, but nature too would be ravaged and scarred. Harnessed in peacetime uses, atomic energy, at least potentially, opened up tremendous reserves of power to mankind that might do much to relieve the pressure of population upon resources. On the whole, however, atomic scientists discounted the easy popular assumption of widespread use of nuclear energy as a source of normal power. Ralph Lapp, in his book *Atoms and People*, warned Americans to conserve petroleum reserves and extract more oil from coal shales. Uranium, too, he pointed out, might become exhausted, although by that time probably solar energy would be utilized. In any event Lapp, like Brown and others, was disturbed by the threat of war and the nagging danger of an increasing population growing faster than its food supply. Thus "humanity quavers before a short-fused superbomb and a slowly ticking population bomb. The forces of the latter are more distant but are no less potent than the former." [31]

XI

POPULATION PROBLEMS

Civilization and technological progress, despite all efforts at the conservation and planned use of resources and energy, were constantly warring against the natural environment. The harmony of man and nature was difficult to achieve, and the problem was complicated even more by the way in which man through his tremendous natural increase further distorted the already precarious balance between population and environment. Not only was modern man the great destroyer of nature with his tools of fire, axe, plow, and arms, but he made matters worse by encouraging population growth. Science and medicine were helping to keep people alive at the same time that the old Malthusian checks of disease, famine, and war were becoming less effective. "All these trends produce overpopulation. If there is to be no present restraint on population, no looking to the future, no doubt that science will always find a substitute for depleted natural resources, then there is no hope of success for those who would conserve the earth's endowment for the future," was one typical modern warning.[1]

Until comparatively recent times world population grew very slowly. Europe at the time of the discovery of America had around a hundred million people, while the American continents were inhabited by only a thin veneer of native Indians. By 1800 the population of European origin, whether still in Europe or in the New World, had about doubled. But in the next century population increased more rapidly, and it continued to do so in the twentieth century. In 1940 the Western world had a population eight times what it had been in 1500, and the ratio of population to land in Europe and America was more dense than it had been in Europe on the eve of Columbus's voyage. Thus by mid-twentieth century the Western world was more crowded than Europe alone had been in 1500.

For the world as a whole, statistics of population growth were even more impressive. In 1650 the world's population was around one half billion persons. By 1800 the total figure was somewhat less than one billion, and a century later, in 1900, more than one and a half billion. In 1950 total population reached two and a half billions for a fivefold increase in the three centuries after 1650. In the last hundred years world population had more than doubled, increasing more in one century than in all previous time. This startling growth, moreover, showed no signs of slackening. Since the Second World War the earth's peoples were multiplying at an annual rate of over one per cent, or between thirty and forty million persons per year. Projected into the future, these figures indicated that the world's population would at least double by the year 2000. Even estimates of a minimal increase gave rise to fears that the next century would see a world in which there was literally only standing room for its inhabitants. "As a matter of arithmetic," a student of population problems pointed out, "it is demonstrable that a population growth of 1 per cent per year could not possibly have been maintained for long in the past;

nor can it continue very far in the future. . . . The projection of a 1 per cent rate of growth into the future gives a population of over 500 billion persons by the year 2500!" [2]

The filling up of the earth's surface, apart from questions of commensurate use or misuse of the land, had grave implications. Short of new frontiers in space, or miraculous developments in science which might drastically revise all previous ratios of people to land, population was dangerously crowding its means of subsistence. Even if famine was averted, the constant increase in population and resultant overcrowding might have disastrous, or at least revolutionary, effects upon man psychologically and emotionally. Man might not prove adaptable to the universal extension and adoption of an urbanized apartment-house type of living. In large cities, Lewis Mumford wrote, "Nature, except in a surviving landscape park, is scarcely to be found near the metropolis; if at all, one must look overhead, at the clouds, the sun, the moon, when they appear through the jutting towers and building blocks." The world's great cities had grown with population and industrialization to meet man's needs, but the mechanical achievement of the city was paralleled by increasing social and physical chaos. Urban populations frequently lacked the most elemental facilities such as fresh air and sunlight, while they faced the results of their disruption of the balance of nature: "ravaged landscapes, disorderly urban districts, pockets of disease, patches of blight, mile upon mile of standardized slums, worming into the outlying areas of big cities, and fusing with their ineffectual suburbs. In short: a general miscarriage and defeat of civilized effort." [3]

The conquest of nature might be a fruitless victory if it created an unnatural environment that man could not happily adjust to or even long survive in. Failing to adapt to his own changed landscape, he might not be able to continue to exist as a species. Consideration of these problems by modern disciples of Thomas Malthus was in marked contrast to the complacency with which eighteenth and nineteenth-century Americans had viewed a growing population. Almost fifty years before Malthus published his famous *Essay on the Principle of Population* in 1798, Benjamin Franklin arrived at a remarkable prediction of the future growth of the American people. In his *Observations Concerning the Increase of Mankind and the Peopling of Countries*, which he published in 1755, Franklin estimated that the population of America would double every twenty-five years and that in a century there would be more Englishmen west of the Atlantic than in the mother country. He also asserted that the migration of peoples did not affect rates of population increase, a view born out by the fact that the rate of increase of United States population was greatest before the large-scale immigration of the late nineteenth century.[4]

In the century after Franklin's death in 1790, when census data could be drawn upon, his prediction of the population doubling every twenty-five years was realized by virtue of the coincidence that the rate of growth was first higher and then lower than Franklin had assumed in 1755. Malthus, who pointed out in his *Essay* that an unchecked population would always increase in geometrical progression, used Franklin's estimates in attempting to ascertain just what the growth of population in American was likely to be. For Americans themselves the crucial question, however, was whether, and for how long, the population of their country could rise at the rate predicted both by Franklin and Malthus.

If conservation was long an un-American idea, even more so was the notion that the United States could ever have a surplus population. As Malthus recognized, the United States was the only country in which population was increasing unhampered by the traditional checks of famine, disease, and war. Throughout the colonial period and well into the nineteenth century, the extensive western lands and rich natural resources of the American continent required a growing population for their exploitation. Malthus' famous doctrine that population tended to reproduce itself beyond the means of its subsistence was almost universally rejected therefore by American thinkers. However valid for the older, mature nations of Europe, it was not thought to have any practical bearing upon the situation or needs of the United States. James Madison, for example, was unusual in his view that Malthus' ideas were inseparable from old countries and awaited the maturing of new ones. Madison was also exceptional among Americans in his opinion that the pressure of population upon subsistence precluded the prospect of banishing evil and achieving Utopia.[5]

More characteristic than Madison's rejection of the idea of progress was the optimism of a number of American economists who, in the early years of the nineteenth century, took issue with the pessimism of Malthus and other English authorities on population. Henry Carey maintained the typical view that the larger the population the greater the progress of civilization. Deprecating the fear that the world's supply of food would eventually be exhausted, he wrote: "Population asks only to be let alone, and it will take care of itself. Without its growth the power of union cannot arise, nor the love of harmony and of peace, essential to the promotion of the growth of wealth and to the cultivation of the best soils, without which the return to labour cannot be large." Among Carey's colleagues there was overwhelming support for the conclusion that, in contrast to England and the Old World, in the United States there was hardly a time before the Civil War when an increase of population would not have been desirable.[6]

The notable change in the historic American rejection of the postulates of the gloomy Malthus, however, did not come until the Second World War. The enormous destructiveness of the war and its tremendous demands upon raw materials put a severe strain upon the world's ability to feed itself. Per capita consumption of food, which declined during the war years, failed to recover and regain even the levels of the economically depressed 1930's. As late as 1952 a United Nations survey indicated that world consumption of food had not yet climbed back to prewar levels.[7] A research report to the United States Congress, based on 1958 United Nations figures, noted that "Statistics are a dry measure of human suffering, but in Asia (excluding mainland China) per capita food production in 1957–58 was 10 percent lower than it was in 1934–38. In Latin America it was 3 percent lower and in the world as a whole it was only 1 percent higher during the same period." [8]

While food production declined, there was no commensurate reduction in population. The millions killed as a result of the fighting and bombing in Europe and the Pacific were less than the continued natural increase in the many parts of the world, including much of Asia, Africa, and America that were untouched by actual hostilities. Even in the midst of world war, new discoveries were made in science and medicine, and improved techniques of sanitation and public health were developed. Frequently this progress encouraged rising birth rates and declining death rates in the very areas where the imbalance between food and population was already greatest. The fact that this imbalance actually increased despite a vast global war served as a grim reminder of the fact that population pressures in countries like Japan had contributed, in the first place, to the coming of the war, and could be regarded again as one of the potential threats to future world peace. The Second World War more than any other single event thus served to revive the warnings of Malthus that population would always press dangerously upon its food supply unless subjected to preventive checks.

In the United States, even though the postwar birth rate soared, there was no lack of food but rather a troublesome surplus of wheat and other crops. For the immediate future food was not a domestic population problem, but the world situation made Americans receptive to the argument that within the century they might also face difficulties in maintaining their high standard of living. The time was, perhaps, not far off when the United States might have to import foodstuffs along with certain strategic raw materials. The example of Great Britain was far from reassuring. As the great industrial nation of the nineteenth century, it had lived upon the raw materials of its empire and upon its world trade. After the war England could no longer subsist profitably by this parasitical technique of unbalancing nature. Thoughtful Americans wondered whether this, too, might not be the eventual fate of the United States, which was playing in the twentieth century the role of Britain in the nineteenth.

Though Americans were a people of plenty, enjoying a period of unprecedented postwar prosperity and economic abundance, the very disparity of their status with that of the rest of the world had its effect upon the American character and concept of nature. The United States seemed to have many of the attributes of what David Riesman called an other-directed society—a bureaucratic managerial society of mass media. This type of society Riesman associated with the stage in history when a nation experienced both low birth and death rates and was therefore entering a period of incipient population decline.[9] At least temporarily, however, the American birth rate rose after the war, and Americans as individuals seemed determined to deny the widespread warnings of the neo-Malthusians in their midst. At the same time, though relatively secure in their own economic abundance, Americans found themselves in an international community shrunken by improved means of communication and agitated by the great disparity in wealth between the industrialized and the so-called underdeveloped nations. The very prosperity and fecundity of the American people served to increase public apprehensions and to contribute to pessimistic fears lest a national wealth not in harmony or balance with nature might not endure.

During the depression of the 1930's the economic contrast of the United States and other areas of the world had not been so marked. Moreover, a declining birth rate in the United States and Western Europe after the First World War allayed neo-Malthusian fears of overpopulation. But the Second World War saw a renewed growth of the birth rate in the Western world, and at the same time an increasing economic gulf between most of the Western nations and the underdeveloped nonindustrial areas of Asia, Africa, and South America. This contrast in the economic position of different areas, plus an ever-growing world population, and the dependence of industrial nations upon overseas imports for their raw materials, all contributed to a resurgence of neo-Malthusian arguments. "One

World" had at least enough reality that one half of the world could not look with equanimity upon the near starvation of the other half. Meanwhile traditional nationalism, with its emphasis upon a larger population and economic self-sufficiency, continued to create an unbalanced production that afforded only temporary solutions even in a wealthy nation like the United States.

The post-First World War alarm over a rising tide of color from the masses of Asiatic peoples was translated after the Second World War into the knowledge that these same masses would not be satisfied indefinitely with their semicolonial impoverished status. Poverty was not new, but there was the new factor of peoples' awareness of poverty, their realization that it was not the inevitable lot of man, and their determination to do something about it. This attitude on the part of the world's masses was often referred to as "the revolution of rising expectations." [10]

Neither able to match the East in numbers, nor to control it economically and politically, the Western world was perhaps becoming more receptive to theories of living in harmony or balance with nature and of stabilizing the population of the globe. The alternative seemed to be to live in an ever-smaller Western island isolated from, and also menaced by, a growing and revolutionary desperation on the part of peoples living on the rest of the earth's surface. With the rich benefits of technology limited to a small part of the world's population, and especially to the six per cent that lived in the United States, the American position was unpleasantly reminiscent of Marie Antoinette's alleged famous advice to the French masses crying for bread: "Let them eat cake!" In the midst of considerable evidence that world living standards were declining, there was the further uncomfortable probability that the United States was not living on a balanced exploitation of its own resources. In practical terms, according to one authority, balance meant that "there can be no hope for a decent life for all mankind unless birth rates in all parts of the world are soon reduced to little more than a third of man's natural birth rate." [11]

Birth control as a solution of the population problem, though radical and impossible of easy adoption, could no longer be dismissed as an extreme or reckless view. For example, one author commented: "We should insist on freedom of contraception as we insist on freedom of the press; it is just as important." [12] The Catholic Church's historic opposition to birth control was becoming less influential, if for no other reason than that the masses of Asia, for example, were outside the church's province. Even in the United States the Catholic position had more political than practical effects. In Japan after the Second World War government dissemination of birth control information helped to reverse the rising Japanese birth rate that had accompanied Westernization after 1867. There were hopes that simple methods of artificial contraception might become usable in other Asiatic nations. Although there was always the danger that nationalistic rivalries and fears of race and national suicide would prevent official encouragement of birth control, the tremendous population increases in most countries seemed to inspire governments to exercise real caution.

In the postwar world birth control as a means of achieving some sort of harmony between man and nature was questioned chiefly by those who saw the possibility of still new frontiers of population in areas like Brazil, or of greater agricultural productivity through the workings of science and industry.

In view of the world's soaring consumption of all its raw materials, it seemed obvious that great scientific and technical ingenuity would be required to feed a growing world population not subject to drastic checks. Agriculture would have to be not only more intensive and efficient, but it would have to tap hitherto unimagined sources of food locked in the atmosphere or the oceans. Good land was constantly shrinking through erosion and urbanization, and the amount of any sort of land was obviously limited by the earth's surface. Even such remaining primitive areas as the interior reaches of Brazil in the upper Amazon River Valley, it was suggested, were already being exploited in an unnecessarily extravagant fashion.

"Brazil, long pictured as covered with thick, green jungle," according to official government reports, "was facing the disasters that follow willful forest destruction—sterile soil, shrinking streams and spreading deserts." Although Brazil held a sizable per cent of the world's forest reserves, these were being consumed for fuel at a rapid rate, or were being burned to make way for agriculture. Subsequent erosion and destruction of water supply and soil fertility indicated that Brazil was being confronted with the same problems of devastation already visited upon the world's more developed regions. Putting more land under cultivation in Brazil, where only a small per cent of the area was farmed, would not only add to the forest destruction but it would probably also increase the existent low population density of the country. In any case some authorities found it difficult to see how areas like central Brazil or the Belgian Congo offered the world any substantial hope of establishing new geographic frontiers. They felt it more likely that these relatively primitive regions would soon duplicate the experience of such troubled areas as Puerto Rico under United States rule.[13]

Puerto Rico was widely pointed to by both neo-Malthusians and their critics as an example of population pressure. One of the latter, Josué de Castro, a Brazilian scholar and authority on food problems who also served with United Nations technical organizations, called Puerto Rico "a very black spot on the map of universal hunger." Although hardly a Utopia under the Spanish before United States annexation, Puerto Rico, "if not exactly swimming in wealth and abundance, was far from the misery and hunger that it suffers in our times." By concentrating land holdings, and by devoting the islands to supplying the American mainland with coffee and sugar, the United States destroyed the subsistence agriculture which had furnished the native population with food. Sanitation and public works, pointed to with pride by the United States, and used as an inducement to new industry, also encouraged population growth and dependence on the United States for food.[14]

For the average Puerto Rican the increased birth rate under the American flag was not accompanied by a corresponding growth in an adequate food supply. Unable to make a decent living in their island home, many Puerto Ricans migrated to the United States, concentrating in the slum areas of New York City, where they helped create new municipal problems. At the same time American magazines carried lavish display advertisements calling attention to the attractiveness of Puerto Rico for industry or for a carefree vacation. What seemed clear was that Puerto Rico was still, despite some economic progress, a land of striking contrasts. The pleasant life of the wealthier minority was based on low taxes and the even lower standard of living of a large majority of the people. Meanwhile, the economy of the island remained sadly out of balance with its environment.

The exchange by large numbers of Puerto Ricans of a rural

slum for an urban one illustrated a particular aspect of the population problem. Not only was world population growing but it was being concentrated in urban centers where it accentuated the lack of balance between man and nature. City life had never lacked critics in America, but flight from the city was no solution to the population problem in the twentieth century. Middle-class suburban areas did not solve any of the major urban problems and, indeed, created new ones. Neither suburb nor city was able to exist in harmony with nature. Although the exchange of goods and services for food and raw materials was to be taken for granted, metropolitan areas used up resources in other less favorable ways. In the city the disposal of man's wastes was an age-old problem still unsolved, although progress was being made in a number of cities in reclaiming sewer waste water. Meanwhile fresh water supplies could be secured in many cities only at increasing expense, and almost nowhere near large American cities were rivers and lakes free of contamination. Even the air breathed was filled with the smoke and fumes of industry and the automobile. Not only was individual health endangered by the soot and smog, but the climate was changed adversely. The one possible exception of higher winter nighttime temperatures in large cities was more than outweighed by the increase in air pollution and in daytime cloudiness, with an accompanying loss of illumination and ultraviolet radiation.[15]

Human welfare was also affected unfavorably by urban life in other ways. The concentration of large masses of people in slum conditions added to the toll of disease and crime, and blighted the lives of countless individuals and families. The cheap labor available to industry was a doubtful compensation for the human misery and social costs of slum areas. Middle-class suburban dwellers, though escaping the city, nevertheless paid a price in higher costs and taxes, and in longer travel time to their jobs. More attention was being paid to city planning, and new technological improvements also promised help in conquering smog and water pollution. But basic urban difficulties could not be resolved as long as the general population continued to concentrate in small areas and to grow everywhere.

Industrialization, often heralded as a cure for the world's population problems, was no step closer to Utopia if it only added still more millions to the world's teeming cities. For the time being, however, most of those who disputed the neo-Malthusians' contentions in regard to overpopulation accepted industrialization as at least a partial solution to food and population imbalances. For example, it was Josué de Castro's main thesis that overpopulation was not the fundamental cause of the world's ills. Present overpopulation, he contended, could only cause real difficulty for future generations. Hunger and malnutrition, and low standards of living, caused overpopulation, while wealthier industrial nations had lower birth rates. Industrialization and better nutrition accordingly were the most immediate steps required to solve population pressures.[16]

Historically, of course, population was curbed in various ways. Starvation and the other Malthusian checks were only the most obvious and effective limitations. But starvation could be potential and operate even in industrial countries of a high standard of living. For example, the straitened circumstances in the British Isles in modern times probably acted as a preventive check even though the threat of starvation was certainly only potential. Immediately, greater industrialization and better agricultural techniques could do something to help overpopulated areas. But in the long run, if the world was to avoid the violence of the Malthusian checks, and if it was to achieve a balance of the forces of man and nature, it would have to exercise an intelligent birth control. In the words of Karl Sax, who challenged Castro's views, "In the future, as in the past, population growth will be controlled by war, famine, and disease—unless birth rates in all parts of the world are soon reduced to moderate levels." [17]

Sax's position was echoed by other writers who revived the warnings of Malthus, but there was sharp division of opinion among experts as to how immediate the population problem really was. In general, English scientists and scholars, living in an area of great population density, seemed more worried than those in the United States. Bertrand Russell, the celebrated English philosopher, asserted that with all the talk of the values of the Western world, the chief one which it could give to the rest of the world was very likely birth control.[18] Sir Charles G. Darwin, the scientist, was probably the most impressive of the British prophets of doom. Sir Charles disagreed with those who believed that birth control or scientific progress could solve the population problem. Any increase in the food supply, he pointed out, would only encourage an accompanying rise in population, and the already precarious struggle for survival would remain. Moreover, the rapid rise of population in modern times foretold the approaching day when even the world's great technological skills would not be able to create enough food. Darwin feared, in short, that mass starvation would be the inevitable outcome.[19]

Some other experts took a more hopeful outlook than the famous English scientist. Britain itself was pointed to as an example of a country in which the birth rate was stabilized, not by starvation, but by the subtle pressures enforced by a lowered standard of living. In a survey on *Population and World Politics*, contributors maintained that world resources were adequate to support a growing population for a number of years in the future. Exhaustion of resources was discounted on the grounds that what was happening was only that *known* deposits of *known* minerals were being depleted.[20] From this point of view technology continued to offer some hope. According to John D. Durand, a population expert connected with the United Nations secretariat, "He who wishes to enter this field of speculation should not ignore the relationship, in the broad sweep of history, between the growth of population and man's ability to turn the resources of nature to his advantage. . . . There may be at least one more step in this evolution, and when that step has been taken, it may appear that the massive growth of population in our time was necessary for it."[21]

In the midst of the popular and scientific attention to the rising tide of world numbers, some concern was also expressed in respect to the quality of a growing population. Some authorities feared deterioration in quality even more than rise in quantity, although most students of eugenics believed that the two problems were really one and the same. Quantity was always difficult to square with quality, and the ever-growing peoples of the world posed a new threat to individuality. The problem, as it was stated by Julian Huxley, was that "In almost all the industrially and socially advanced countries, the level of innate intelligence, and probably of other desirable genetic qualities, is decreasing generation by generation." The reason for this, most eugenicists believed, was the differential birth rate, or the fact that the lower and poorer classes of the population had more children, just as the underdeveloped areas of the world had higher rates of population increase. Efforts on a world scale to provide equality of opportunity among all reaches of the population, it was sometimes feared, might harbor the seeds of biological disaster. If the differential birth rate was maintained, economic aid and welfare measures would enable the less fit to receive more opportunity for advancement and reproduction. On the other hand, the continuance of impoverished environments encouraged a retarded development and an excessive fertility which, in turn, made more difficult all economic and social betterment.[22]

Population problems were essentially matters of large numbers, involving a balance between aggregate resources and peoples. But real harmony also meant individual adjustment. Whatever the merits of the arguments of the eugenicists, their point of view called attention to individual standards of quality in the midst of the more general concern over population *en masse*. In a sense, too, the harmony of man and nature over the future would depend on the kind of philosophy of life worked out between individuals and a machine society.

XII

THE INDIVIDUAL AND THE MACHINE

In the twentieth century the problem of man and nature and the conflicting philosophies of harmony or progress were often expressed in terms of the juxtaposition of the individual and the machine. The conservation movement was also being broadened to include human as well as natural resources because man, too, was a victim of industrial civilization and its blighting effects upon the land. Man was endangered even more by the machine world he had created than by his environment. Siegfried Giedion, in his comprehensive book *Mechanization Takes Command*, observed at the outset of his inquiry that he was motivated by "the desire to understand the effects of mechanization upon the human being: to discern how far mechanization corresponds with and to what extent it contradicts the unalterable laws of human nature." [1] Arnold Toynbee suggested that technological change had already speeded up beyond the individual's ability to absorb or cope with it. As the machine moved at a rate faster than human nature could go, man was in danger of becoming as out of harmony with his times as the primitive Indians. "I think," Toynbee said, "the effect of our vastly accelerated pace of technological progress has been that we have now made ourselves into our own Pueblo Indians and our own Aztecs." [2]

It seemed true that if modern man was to achieve equilibrium with the forces of nature, he would have to work out a *modus vivendi* between the individual and the machine. Not only was modern technological progress achieving constantly greater miracles in changing and putting to use the natural environment, but the god of the machine was threatening to dominate man himself. Thus the concern of scientists and economists over depleted natural resources and a multiplying population was matched by an equal solicitude on the part of humanist philosophers over the preservation of the individuality of man. In other words, in the midst of popular attention to what man was doing to his environment by transforming the face of the earth, there was also reason to look carefully at the way in which individuals were having to live in a world increasingly controlled by science and technology. Conservation, therefore, might properly include what a machine-made environment was doing to man.

Susanne Langer, a distinguished philosopher, feared that technical progress was threatening man's freedom of mind. In adjusting to the modern world's technological society, man was in danger of losing all contact with the natural world. "The ordinary city-dweller," she pointed out, "knows nothing of the earth's productivity; he does not know the sunrise and rarely notices when the sun sets; ask him what phase the moon is in, or when the tide in the harbor is high, or even how high the average tide runs, and likely as not he cannot answer you. Seedtime and harvest time are nothing to him. If he has never witnessed an earthquake, a great flood or a hurricane, he probably does not feel the power of nature as a reality surrounding his life at all. . . . Nature, as man has always known it, he knows no more." [3]

The feasibility or desirability of attaining some sort of balance between man and nature involved important ethical and practical considerations. On the one hand, man's ever-growing population and use of natural resources seemed to point to the need of conservation and careful planning. On the other hand, it was also true that man had been helped in gaining his high estate through his technological ingenuity in remaking and exploiting his environment. But as the pace of progress speeded up in the twentieth century, more observers began to wonder how long such constant acceleration might continue before it produced a universal crack-up. This was a practical question for the scientists and experts, and it was also a matter of ethics. Poets and philosophers might well inquire whether, assuming its ability to hold on, the progress of such a machine-made civilization did not exact too high a price in terms of human values. Until the advent of modern man the world had exhibited a relatively steady state of balance and harmony in nature. Now, in the second half of the twentieth century, it was appropriate to ask whether the sustained exploitation of nature by the powers of science and machinery was really desirable in the long run.

Back in 1861, when the United States was facing the political crisis of the Civil War and was entering upon the main phase of the industrial revolution, one of the last of the spiritual descendants of Jeffersonian individualism tried to build his fortune on the California frontier. In the land of golden promise young Henry George found himself baffled by the contrast between the average individual's poverty and the rich natural abundance of the environment. To his sister back home in the East he wrote: "Sometimes I feel sick of the fierce struggle of our highly-civilized life, and think I would like to get away from cities and business, with their jostlings and strainings and cares, altogether, and find some place on one of the hill-sides which look so dim and blue in the distance, where I could gather those I love, and live content with what Nature and our own resources would furnish; but alas, money, money is wanted even for that." [4]

George discovered that technological progress was not an automatic solution to poverty. In an interesting article entitled "What the Railroad Will Bring Us," he speculated on the effects of the approaching completion of the first transcontinental line to California. Local property owners eagerly envisaged a coming increase in land values, but George, already anticipating his famous criticism of the unearned increment in land valuation, lamented that California had not been settled on the basis of small free homesteads. "The locomotive," he warned, "is a great centralizer," which would kill small towns and small business and make possible large fortunes and also a poor class. "We need not look far from the palace to find the hovel. . . . Amid all our rejoicing and all our gratulation let us see clearly whither we are tending. Increase in population and in wealth past a certain point means simply an approximation to the condition of older countries—the Eastern states and Europe." [5]

Henry George's celebrated attack on real estate speculation and his advocacy of a tax on the unearned increment that came to such ownership as a result of a rising population were an agrarian's plea for a better balance between man and nature. Curbs on the private exploitation of the natural resources of the land would make possible a more equalitarian division of property without the collectivism and statism of a planned socialist society. Man would again be humanized and freed of dependence upon the machine. Unlike conservationists, who wanted to make sure that the government controlled the land, George wished to entrust it to the people so that they could use it. With his concern over poverty in the midst of natural abundance, Henry George was not a precursor of the conservationists who worried more about the exhaustion of diminishing resources than about the distribution of the natural surplus. Instead, *Progress and Poverty* was an inspiration for the later humanists and agrarians who sought in an age of exploitation to defend the individual against the machine.[6]

Henry George's philosophy was a bridge between the ideas of the Jeffersonians and transcendentalists and the point of view of the modern agrarians and humanists. George's own thinking had been stimulated by observation of the social and economic disruption that followed in the wake of alternating cycles of prosperity, depression, and war. A half century later another generation was profoundly stirred by the impact of two world wars and intervening decades of prosperity and depression. As the level of the mass production and consumption of material goods rose, the economic prosperity, sense of power, and leisure time of the average individual seemed to increase. But there was another side to the story. In the midst of the good times of the 1920's, a small minority of critics questioned the new materialist values that had come to depend upon machine production. The machine, while increasing human power, had also decreased the individual's older craft skills and his contact with nature. Mass production made men live to consume, as the constant rolling of the assembly line became an end in itself, and as men desired more and more gadgets for their comfort and recreation. It was also pointed out that the notion that man's welfare depended on an unending increase in production was destroying the resources of the earth and the time man had to enjoy them.

These attacks on industrialism and the machine, at the very climax of their seeming success in the twenties, indicated that technology failed to satisfy certain basic human needs. According to the manifesto of a group of humanist writers and scholars in the South, man's labor and consumption had been hurried and brutalized by a machine civilization. In taking a stand for a national agrarian movement, these Southerners went beyond economics. Their volume, first published in 1930, was also a book about nature, and of man's relationship to nature.[7]

In the depths of the depression of the thirties, Lewis Mumford indicted the machine for failing to provide properly for human activity. "Western society," he asserted in *Technics and Civilization*, "is relapsing at critical points into pre-civilized modes of thought, feeling, and action because it has acquiesced too easily in the dehumanization of society through capitalist exploitation and military conquest." Mumford was doubtful whether machines would have developed so rapidly without the extra incentives of commercial profits and war. This debt to capitalism and war he believed was unfortunate because it militated against the use of technics for social welfare. Although the dream of conquering nature was one of the oldest in man's mind, Mumford now saw the ultimate goal as going beyond the mere conquest of nature to that of a constant resynthesis—steam for horsepower, rayon for silk, etc. Invention had become a duty and "To live was to work; what other life indeed do machines know? In short, the machine came into our civilization, not to save man from the servitude to ignoble forms of work, but to make more widely possible the servitude to ignoble standards of consumption that had grown up among military aristocracies." [8]

Mumford's critique recalled the pre-Civil War transcendentalists' skepticism of American inventions, and their complaint that the individual was being swallowed up by the machine. Thorstein Veblen, as Mumford noted, had also developed the paradox of technical progress. The author of *The Theory of the Leisure Class* and of *The Theory of Business Enterprise,* in his later volume *The Instinct of Workmanship and the State of the Industrial Arts* attacked the usefulness of modern invention. While necessity was often the mother of invention, it was also true that obsolescence was encouraged by business enterprise and its advertising techniques. Some modern inventions therefore met no wants other than those that they themselves created. Although the telephone, typewriter, and automobile were great and useful achievements, Veblen nonetheless was "at least doubtful if these inventions have not wasted more effort and substance than they have saved,—that they are to be credited with an appreciable net loss." [9] Also pertinent to Mumford and Veblen's skepticism was Bertrand Russell's wry observation that each improvement in locomotion had increased the area in which people were compelled to move.[10]

Mumford, more than any of his mentors or fellow critics, in his analysis of the machine called for a new philosophy of balance and harmony. Possible answers to the superfluous power of a purposeless materialism were the cult of the past and a return to nature. These extremes, he believed, could be averted if man would economize production and normalize consumption. Since vital wants were limited, stylistic obsolescence and the dogma of increasing wants had to be cultivated so that production, not for needs but for mere acquisition, could be continued. Such a rapid turnover in consumption, however, tended to destroy the laborsaving gains of mechanization and technics. Laborsaving, Mumford maintained, could take place only when standards of consumption remained relatively stable. Otherwise man was chained to a treadmill, producing and consuming more than he needed in order to keep his place in the race. For a way out of the dilemma, he offered his concept of a dynamic equilibrium, of balance rather than rapid one-sided advance, and of conservation rather than reckless pillage. Equilibrium in the environment would restore harmony between man and nature, between industry and agriculture, and between man and the pressures of overpopulation.[11]

Mumford's concept of a dynamic equilibrium temporarily found some answer in the efforts of the New Deal to balance production and consumption. The depression furnished a congenial climate for acceptance of Mumford's point that the lure of commercial profits had stimulated the creation of machinery and a resultant production that bore little relationship to social needs. Profits, technics, overproduction, collectivism embraced a cycle that reached its climax in the collapse of the 1930's. Production as an end in itself had proved self-defeating because production emphasized machine technics, and the machine pointed towards a collectivist society which would destroy capitalism. This was the thesis of Erich Kahler's book, *Man the Measure: A New Approach to History*, published in the mid-war waning days of the New Deal.[12]

Kahler's belief that individuality was doomed by the conquering march of a technological collectivism continued to dominate economic thinking after the war. But economists, under the spell of postwar inflation and cold war rivalries, found it difficult to reject the goal of an ever-greater production of material goods. For those willing to give up the vestiges of traditional free enterprise, John K. Galbraith, professor of economics at Harvard and a popular author, offered a solution. In answer to the crucial problem of values implied in the question "How Much Should a Country Consume?" Galbraith suggested that, if conservation was deemed important, then government controls to cut down the production of unneeded goods were warranted. But since full employment was desirable, there should be a parallel increase in the social, cultural, and recreational services that had high labor, but small materials, requirements. Galbraith dismissed the view that uninhibited consumption was necessary to preserve traditional individual liberties. Though critical of the private conspicuous consumption of goods and resources, he did not question the government's own spending, or the high level of economic waste associated with the military defense program.[13]

In general economists, in their concern over economic and technological efficiency, were inclined to overlook the interrelated roles of man and nature. Galbraith, at least, went beyond most of his colleagues in suggesting a means of pairing both conservation and continued economic growth. After the Second World War, as man continued to be torn between the forces of natural harmony and technological expansion, economic dogmatism was displaced in part by a more humanistic concern over the implications of the new age of science.

Lawrence K. Frank, author of *Nature and Human Nature*, was one of the philosophers of science who set the tone for the more modest view of man's capabilities that was a characteristic of the thinking of the 1950's. "We humans like to think of ourselves," he wrote, "as the most important of all organisms, as the stars in the play, with the geographical environment and all the plants and animals therein as scenery and props for our human drama, if and when we do think of them. But this self-centered view of nature and ourselves, while often comforting, is both misleading and obstructive. We should try to think of ourselves as one of many organisms on this vast stage of nature, where we must recognize that we play only one part in this ongoing drama of existence. Usually we are a disturbing, often a destructive, agency, but like other organisms, we are involved in the complicated interrelationships and we actively participate in and help to maintain the totality we call nature. Indeed, we should think of ourselves as being used by other organisms and natural processes, one of the many configurations of energy that make up the totality of existence, partaking in and being carried along by these larger equilibrating and compensating processes of nature." Frank emphasized that the human fondness of speaking of man's ability to control nature "means that as man learns through scientific study to understand the order and processes of nature, he must increasingly 'obey nature,' that is, he must think and act in accordance with natural processes. He can control nature and get what he wants only by learning to think, act and work according to the requirements of nature." [14]

In line with Frank's point of view, Samuel H. Ordway's *Resources and the American Dream* stated "the problem of the expanding industrial consumption of resources." Ordway's thesis was that this continued use could produce scarcities serious enough to destroy the American standard of living and culture. A theory of limited growth, he argued, would not necessarily mean stagnation or decay if consumption was gradually cut down so that supply was not exhausted and so that

Americans ceased living upon their resources capital. In urging his solution of a balanced civilization, Ordway complained that "Success has led us to believe that the earth is a cornucopia, and the machine a god. It has led us to a false faith in man's omnipotence." If free enterprise was to avoid government regulation it would have to balance its resources budget. "We can," he concluded, "hope and work for continuing growth as our technologists deliver on their promises, but growth itself is less vital than stability with freedom." [15]

Some scientists hopefully saw man moving to a new stage of social responsibility in his use of nature. Thus at the Princeton symposium on "Man's Role in Changing the Face of the Earth," E. A. Gutkind, an authority on urban planning in England, in a challenging essay on "Our World from the Air: Conflict and Adaptation," observed both "the general mess" and the "vast parts of the earth's surface still unused." Man, he pointed out, first tried to protect himself from his environment in his villages, then in a stage of growing confidence he began to reshape the environment. This led to the third or present stage of aggressiveness and conquest in which adjustment yielded to the neglect and exploitation of nature. Gutkind saw finally a fourth stage of responsibility and unification in which interaction between man and his environment would increase. Man would also move into a new relationship with his world or cosmos until he would no longer live in a man-centered or earth-centered universe.[16]

Before such visions could be realized, man had yet to overcome the cultural lag in his thinking, or what Sir Charles G. Darwin called "The Time Scale in Human Affairs." Human experiments in the use of the environment, he pointed out, were all necessarily subjective and governed by the life span, or rather the adult working life span, of a generation. This, he believed, was the most fundamental limiting difficulty in planning about humanity. His life span tended to make man conservative about long-range changes. Technology might be radical, but even in a scientific age crafts remained, and these tended to be conservative. Thus politics and agriculture were crafts wedded to older ways, while political ideas, for example, were framed by men in their twenties but were seldom applied until their authors were in their fifties and had reached high office.[17]

Like Darwin, Carl O. Sauer, a distinguished American soil scientist, was amazed at the conservative lag in both the popular and the scientific consideration of the need for conservation. Although Sauer was critical of certain conservationist techniques which he believed would be beneficial for only a short time, he was uncompromising in his attack upon the modern world's increasing emphasis upon greater production and consumption. Led by the example of the destruction of two world wars and the physical scientists' faith in technology, material progress and the capacity to produce and consume had become the watchwords of modern times. Yet, he pointed out, what man had largely learned was "how to deplete more rapidly the resources known to be accessible to us. Must we not admit," he added, "that very much of what we call production is extraction?" Since the time of Columbus, European expansion had been a mixture of self-interested exploitation plus a civilizing mission. The latter was redefined from time to time until now it was called helping the underdeveloped nations. Industrialization of these nations, however, was mainly to secure benefits to the Western world, while the labor of the so-called underdeveloped peoples was increased. In the same way the introduction of advanced techniques of agriculture to increase the local food supply and general level of productivity might in the long run not be beneficial to the native lands.[18]

Although living beyond one's means was now the fashion, Sauer ventured to predict that people might tire of getting and spending as a way of life. They might resent the loss of individuality and increasing government controls which had accompanied the drive for a greater productive output. Perhaps what was needed, he felt, was a return to an older ethics and aesthetics of prudence and moderation. There was the danger, too, as another scientist asserted, that, in the course of modern evolution and history, man might "ultimately suffer the loss of that earlier wisdom of his non-technological civilizations which sustained and nourished his higher values and his more intuitive relations to nature." [19]

Scientific concern for the preservation of a traditional individuality and a proper balance or harmony of man and nature ran up against the realities of continued national rivalries. Cold war competition between the United States and Soviet Russia made it difficult to think in terms of "One World" even when the requirements of a coming space age seemed to put a premium on global unity to meet the challenge of the cosmos. The most pressing immediate problems, however, continued to be those of man and his earthly environment. Both America and Russia concentrated tremendous energy upon the struggle to dominate and exploit the remaining underdeveloped areas of the world. The idea of a balance of nature, in which man recognized the limitations of natural environment beyond which it was not wise to go at a given moment, was hostile to "the Soviet purpose to use science as a means of mastering all environmental limitations." Lysenko and the Soviet biologists, in resurrecting the Lamarckian theory of the inheritance of acquired characteristics, attempted to provide scientific support for an extreme Russian environmentalist position. By favorably changing the environment individual man might be improved, and, according to Lamarck and Lysenko, this progress transmitted biologically to succeeding generations. The official Soviet view of the conflict between heredity and environment reverted in a sense to the extreme optimism of the eighteenth-century believers in unlimited progress. The Soviets, like reformers in the pre-Darwinian scientific world, were confident that the environment could be reshaped to insure perpetual progress.[20]

Although the harsh Russian environment and the ideology of the Communist Revolution both placed a premium on the mastery of nature, the American emphasis upon machine production exemplified the same majestic lack of concern for any concept of balance. In the cold war competition with the Soviets, American experts stressed the importance of outproducing the Soviet nation and utilizing to the hilt America's superior mass technology. Export of this technology was widely regarded as the best weapon in winning the allegiance of the less developed areas and nations of Asia, Africa, and Latin America. Independent countries naturally attempted to extract from Russia or the United States as much economic and technological assistance as possible, holding out the reward of ideological and military support in the cold war.

Though defended most often in terms of the national interest, the United States foreign aid program was also justified as the best means of raising world economic levels to the American standard of abundance. Americans believed, too, that their values of individuality and democracy depended on the prior achievement of a satisfactory level of economic prosperity and productivity, and they assumed accordingly that mass production and consumption were necessary and desirable world goals. The American people naturally resented criticism of the foreign aid program as an imperialist device, although by their own definition of its terms, it seemed clearly to point to the Americanization of the world. The real problem, however, was not one of imperialism but of balance. In terms of the world's need to achieve a more harmonious relationship between man and the earth's resources, the American way of life could be regarded as a disrupting factor. Frequently it upset rather than restored the economy of underdeveloped areas. According to the European author of *The World the Dollar Built*, "What the world needed and still needs most from America is that she put her own house in order, that her giant economy be made stable, predictable, expanding, and able to trade and co-operate fully and consistently with other nations." [21]

A responsible American critic pointed out that "Our productive and consuming capacity is greater than that of any other country in the world. But the result is that our principal standards are standards of quantity: we have more of everything than anybody else—automobiles, refrigerators, radios, railroads. Consequently, our ideal is beginning to be not so much a world peopled by wise and happy men as it is a world in which 'every family had its automobile and every pot its chicken.' We have too easily made the assumption that other values would automatically follow our material well-being, that out of our assembly lines the good life would spontaneously be born." [22]

The idea that there was some direct relationship between programs of economic aid on the one hand and political attitudes on the other was also subjected to critical scrutiny by George Kennan, an important co-author of American postwar containment policy. Kennan questioned at the start the "absolute value attached to rapid economic development. Why all the urgency," he asked? "It can well be argued that the pace of change is no less important than its nature, and that great damage can be done by altering too rapidly the sociological and cultural structure of any society, even where these alterations may be desirable in themselves. In many instances one would also like to know how this economic progress is to be related to the staggering population growth with which it is associated. Finally, many of us in America have seen too much of the incidental effects of industrialization and urbanization to be convinced that these things are absolute answers to problems anywhere, or that they could be worth *any* sacrifice to obtain."[23]

Kennan's query concerning the Soviet-American competition to exploit or aid the remaining underdeveloped regions of the world brought into proper focus the exaggerated demands exacted of man and nature by the cold war. People remained too conservative to take a global view of what were really world-wide problems; yet at the same time they were just progressive enough to jeopardize older more primitive virtues. Although man dominated his world, his mastery was far from complete. He was still subject to many ungovernable environmental controls, and the natural processes of the earth had not ceased to operate simply because man existed. Life itself, the scientists reminded their fellows, had been present on the earth's surface for only a small fraction of global time. If man was to break away from the past and start to control his evolutionary process, the results might be such that his mental ability would outstrip his physical or psychic ability to survive. Or, perhaps, man would achieve "a domineering brain which so completely controls environmental processes that its own physical requirements become negligible." Biologists, however, warned that even if man escaped genetic damage from radiation, or a more direct and immediate nuclear damage, he might not be able to survive the evolution of pathogenic organisms like those which caused the Black Death in the fourteenth century.[24]

In the midst of the mixed counsels of optimism and despair, a temperate statement of "The Process of Environmental Change by Man" was offered by Paul B. Sears, chairman of the Yale University Conservation Program and a past president of the American Association for the Advancement of Science. Sears pointed out that, with all his flexibility as an organism, man was still dependent on the resources of his environment. Man, he added, "is clearly the beneficiary of a very special environment which has been a great while in the making. This environment is more than an inert stockroom. It is an active system, a pattern, and a process as well. Its value can be threatened by disruption no less than by depletion."[25]

The Christian and Judaic tradition set man apart from nature in a dualistic relationship, and this view, Sears believed, was shared by "those who resent, for whatever reasons, any warning sign along the road to a perpetually expanding economy." Although some scientists surprisingly joined in this view, Sears felt that "Mankind is not well served either by hysteria or by false visions." While it was true that man was ingenious and that America still enjoyed great resources, all approaches to the future, except the most extreme technological one, "indicate that humanity should strive toward a condition of equilibrium with its environment." With all the customary environmental changes produced by man, change finally became a problem of ethics. Sears concluded therefore that "Whether we consider ethics to be enlightened self-interest, the greatest good for the greatest number, ultimate good rather than present benefit, or Schweitzer's reverence for life, man's obligation toward environment is equally clear." [26]

The difference between modern man's worship of the machine and the ancient dream of universal harmony was well summarized in an address by Professor Sears which he entitled "The Steady State: Physical Law and Moral Choice." Though Sears warned that the goal of balance and harmony was an elusive one, he spoke for those who believed that the achievement of a dynamic equilibrium between man and nature remained civilization's greatest problem and standing challenge for the future. "The infinite variety and beauty of the world about us, the incalculable facets of human experience, the challenge of the unknown that must grow rather than diminish as man advances in stature and becomes at home here—these," he asserted, "are sufficient guarantee that a stable world need never be a stagnant one." [27] Sears did not believe that there was any conflict between individuality or the human adventure and "a respect for the order of nature." Instead, continuance of the human adventure "for so long and at as high a quality as possible" required this respect. In contrast, an ever-expanding economy had no physical warrant in nature. Finally, in his plea for an ethics and aesthetics in nature as well as for a harmony and balance, Sears suggested that the only tolerable landscapes were those untouched by man and the ones where he had achieved a balance with nature.[28]

Among the most persuasive modern exponents of an equilibrium between the forces of man and nature was Joseph Wood Krutch, an ardent defender of the individual and critic of the machine. Professor of English at Columbia University until 1952 and a noted drama critic, Krutch left New York City to live in the Southwest, where he could be closer to nature. Earlier in his career he had published a biography of Thoreau, and there was much of the famous transcendentalist in his own essays and point of view. Recalling Thoreau's advice to "Simplify," Krutch indicted modern advertising and technology for their pressure upon the American public to engage in a wasteful and ostentatious consumption. Material standards, which defined the good life solely in terms of an ever-greater production, and the prospect that scientists might fathom new ways to feed more and more people, merely threatened what little hope there still was for retaining some of nature's natural wildness. Krutch distrusted the social scientists' bland assurances that man was learning to control both nature and human nature. Like Thoreau, he believed that nature was the best teacher, and he urged the public to consider its conservation particularly in terms of mankind's nonmaterial needs. "Unless we think of intangible values as no less important than material resources, unless we are willing to say that man's need of and right to what the parks and wilderness provide are as fundamental as any of his material needs, they are lost." [29]

Only the United States among highly developed nations offered its citizens the opportunity to visit large regions where nature still dominated the scene, but the "variety of nature grows less and less. The monotony of the chain store begins to dominate more and more completely. One must go farther and farther to find a window in which anything not found elsewhere is to be seen." [30] Even though the density of population remained low in many parts of the Western United States, the automobile was able to carry the environment of the city to the country. Rural areas were made to conform to the national pattern of life, and their inhabitants lost their local pride and love of nature. Scenic parks were in danger of becoming wilderness slums as a result of the crush of casual visitors brought by the automobile.

In answer to the argument that modern civilization needed more and more land, Krutch admitted that probably most school and city park sites would bring a higher economic return if used for factories or office buildings. However, a growing population had not yet reached the point where every scrap of land had to be put to its so-called highest and best use. Krutch continued to believe that the wilderness and idea of the wilderness was one of the permanent homes of the human spirit. But, he wrote, "If desire for contact with nature and some sense of unity and sympathy with her are merely vestigial hankerings surviving from the time when man lived in a more primitive culture; if these vestiges can, and should, fade gradually away as he becomes more and more completely adapted to a civilization founded upon technology rather than upon natural processes—then obviously there is not much point in trying to preserve opportunities for gratifying the hankering." Krutch was little comforted by the realization that the unspoiled environment of the American Southwest would probably last for his time.[31]

Although the world grew constantly more crowded, people continued to exploit the resources of what they believed was their earth, with little concern for the future. Even in the Southwest, in the rugged region of the Grand Canyon of the Colorado River, it was difficult to escape the forces of technology. The river had been dammed, someday the canyon would probably be bridged, and at Los Alamos nearby the first atomic bomb had been made. The individual and nature were both succumbing to the seductive power of the machine. But survival in the future might depend less on science than on the world's political ability to prevent a nuclear war.

XIII

CIVILIZATION AND WAR

Modern war stood as the greatest single challenge to the idea of a balance or harmony between man and nature. Formerly only an internecine struggle of the human race, since the discovery of the atomic bomb it affected the whole natural world. Not only did the prospect of thermonuclear warfare threaten the survival of mankind, but the ensuing radioactive fallout could damage the physical earth to the extent that all forms of biological life might be impossible for years to come. Perhaps this was the way other long-forgotten earlier civilizations had died, their finite problems melted away in the crucible of a vast cosmic catastrophe. At least dissolution would insure a fresh start and a slow rebuilding in the long evolutionary process from simplicity to complexity. Such a view, however, was small comfort to man circumscribed as he was by space and time, and with an historic angle of vision little wider than the perspective afforded by his own generation. After all, as Arnold Toynbee pointed out, despite two world wars and the "haunting fear of the advent of a third," Western man was still not cured of his egoistic beliefs that he was not as other men are and that his civilization could escape the fate of past ones.[1] Although war had always been a great waster of life and resources, man had never before, so far as he knew, faced consciously the awful prospect of the complete and sudden destruction of his civilization.

To avoid catastrophe was no easy task, even though the fact that man had survived until now seemed the best argument for his future well-being. Prediction on the basis of historical trends was also hazardous, but if one accepted in any way the desirability or necessity of achieving some sort of balance in the relationship of man with his environment, the tendencies of the Western world were profoundly disturbing. The very technological success and aggressive expansionism of Western society encouraged the economic and political conflict of nations. Perhaps the industrial revolution had stimulated the West to undertake more than it could manage. In exporting its total war and technological progress, the West had made possible its own destruction. Meanwhile, the ever-accelerating tempo of change seemed to insure that the eventual crack-up would not be long deferred. Although the death of a civilization could come in various ways, through overpopulation or from some malfunctioning of the human species, the most pressing immediate concern was the possible outbreak of another world war.

In the twentieth century war had become the great common denominator of modern civilization. As Charles and Mary Beard wrote in the midst of the Second World War, "Despite the mutability of things human, there is one invariable in the history of men and women. This is war. And inasmuch as the efficiency of war in spreading death and destruction depends upon some degree of civilization, it follows that, subject to the law of thermo-dynamics, if there be one, the future of civilization in the United States has at least this much assurance." [2]

Whether the Beards were correct in their gloomy conclusion that war was the chief continuing evidence of some semblance of civilization, there was little doubt that it was a mainspring of modern life. The nation-state, probably the most important institution in the twentieth century, was largely a mechanism for the preparation and waging of war. During the First World War, Randolph Bourne had declared with sardonic bitterness, "War is the health of the state." Looking back later upon two world wars, George Santayana, the expatriate American philosopher, observed that government was "a modification of war, a means of using compulsion without shedding so much blood." [3] Though it was a truism that governments rested in the last analysis upon force, it had also become worthy of note that the ever-present shadow of modern war made it easier for governments to exercise their compulsive powers. Peoples everywhere were cowed by talk of national security, even though national security no longer had any real meaning in terms of the life of the individual or the conservation of his environment.

Nuclear destruction added new dimensions to the ways in which science and technology contributed to the making of war by the state. Long before the detonation of the first atomic bomb, war had already become intimately related to industrial civilization. If war was the health of the state it was also a most useful adjunct to that other important institution of modern civilization—an industrial society. War and industry were allies in their need for mass organization and centralized government. Both sought standardization and accelerated production. War also encouraged industrial activity through the destruction of its products, either by use or obsolescence. Mass production techniques in science and industry were largely responsible for modern war becoming total in scope, affecting the entire population and not just the fighting forces. This enlarged role of modern total war in turn augmented the authority of the government. War, actual or potential, therefore seemed a central feature of the dominant political and economic institutions of the age.

The agrarian statecraft of early American history, with its philosophy of limited government, had been restrained in its appetites. To a considerable extent it accepted the notion that there must be some necessary balance between man and nature. In contrast, modern industrial statecraft had no real philosophy for civilization apart from war. Under the industrialist statecraft of the twentieth century, ends and means were confused. War was no longer a last resort for achieving limited political and economic goals; instead the resources and energy of man and nature were constantly directed toward war. Multiplying functions of government, increasing productivity of industry, the quest for social and economic security, and the search for new markets were all intimately involved with preparedness for war. In a world so obsessed neither individual or nature had any intrinsic value. The destruction of resources and of masses of men was accepted equally and alike as a necessary sacrifice to Mars. The slogan *c'est la guerre* had never carried so much meaning.

War in itself seldom provided solutions to world problems. And a modern nuclear struggle, which could only be self-defeating, did not promise to gain even the temporary protection historically associated with past wars. Even if nuclear war were never actually fought, continued preparedness for it and its ever-present threat were calculated to destroy any prospect of balance or harmony in the life of the planet. Another world war could wreck civilization, not only by a sudden explosion, but through the corroding cancer of mounting human fears and the unceasing exploitation of all remaining natural resources.

There was no doubt of the influence of military needs and thinking upon the way of life in the United States and other modern nations, but the new militarism often escaped attention because it differed so considerably from its historic predecessors. Militarism in its older form of standing armies on parade had been succeeded by such developments as the garrison state and a permanent war economy. As the power of the military extended into areas of American life previously reserved for civilians, the contrast between the citizen and the soldier became less apparent. In other words, military considerations overshadowed normal civilian requirements to the point where politicians became the servants or allies of military leaders, overwhelmed by their prestige and confused by the technicalities of defense in modern war. With virtually the entire population enrolled in some military capacity, or dependent in an economic sense upon military spending, the traditional conflict of the military and the civilian was minimized by the gross subordination of the latter. Thus, as one modern writer put it, it was "possible that, in the United States as elsewhere, the technological implications of modern warfare may make possible a new type of militarism unrecognizable to those who look for its historical characteristics. . . . Anyone who thinks for one moment of the effort involved in building the atomic bomb will not find it difficult to realize that, in the new warfare, the engineering factory is a unit of the army, and the worker may be in uniform without being aware of it. The new militarism may clothe itself in civilian uniform; and, if the present relations of production are maintained, it may be imposed upon a people who see in its development no more than a way to full employment."[4]

In the United States the new militarism was most apparent in the large percentage of the national budget devoted to war or defense, and in the close connection between the armed forces and American industry. After the Second World War, to support the largest peacetime military establishment in American history, approximately one third of the Federal budget, or some $12 billion, was appropriated in 1947 for the armed forces. With the Korean War the military budget jumped to three and four times the 1947 figure and throughout the 1950's averaged between $40 and $50 billions annually. In the Federal budget for the fiscal year 1960, more than one half of total expeditures was allocated for military functions, overseas military assistance, atomic energy, and stockpiling vital raw materials. If veterans' payments, interest on the national debt, and other national security categories were added, over three quarters of the Federal budget could be considered as war-related. In any case the Defense Department was being allotted in a single year more than the total of what it had received in all American history up to the First World War, and the per capita cost of military security since that earlier conflict had multiplied more than one hundred times. Moreover, since many of the items in the defense budget were now of a continuing, semi-permanent nature, Congress was tending to lose control of its constitutional power to make appropriations. The military, as the largest part of the Federal bureaucracy, was virtually an autonomous and self-perpetuating vested interest.[5]

It was obvious that defense expenditures so vast and so rigid could not help but give the military a commensurate influence over the American economy. Whatever their own desires and ambitions, top-ranking officers gravitated to positions of power in private industry as well as within the government. Procurement for the armed services and the development of atomic energy was bigger business than American industrialists had ever envisaged. Since the Korean War, economists estimated that one third of the nation's business was coming from defense spending, and the armed services' investment in plant facilities such as bases, equipment, and supply depots exceeded by far the assets of America's largest corporations. In the field of scientific research military needs were dominant, and pure research commanded less and less support even in the nation's colleges and universities. Instead of the industrial arts supplanting the military, which was a fond liberal dream of the nineteenth century, the twentieth century saw their coalescence. The Industrial College of the Armed Forces, a top-level service school, illustrated the close liaison between the armed forces and American industry. At the college representatives of industry mingled with officers and civil service personnel, while the students took educational field trips to important industrial plants. Industry also cooperated with the Defense Department in securing needed appropriations from Congress. The business of industry was national defense, and national defense had become big business.[6]

The rise of military power and influence after the Second World War aroused fears that America was becoming a garrison state, with all the more humane and civilized values in life subordinated to war-related activities. Certainly preparation for war was becoming more expensive. The tremendous rate of world rearmament could not continue indefinitely without seriously diminishing the world's capital stock of natural resources and without adding to the possibility of undemocratic bureaucratic controls. All this was perhaps less apparent in the United States, where the nation's historic wealth made it possible to have guns and butter—or at least oleomargarine. But some American economists believed that war could no longer, as in the past, be financed by increased production and currency inflation. In the future, if the United States was to continue on a permanent war footing, there would have to be more actual savings and conservation. This could come only in the form of declines in civilian consumption and the standard of living, enforced by a program of higher taxes. As the competition between civilian and military needs grew, more and more government control over the economy would follow, and the free play of the market would disappear.[7]

Even if the high rate of military spending fastened on the world by the cold war was feasible, at least temporarily, from an economic standpoint, it was open to strong moral objection. Extravagance and corruption were encouraged by the sense of haste and urgency. Because of the quick technological obsolescence of so much military equipment, waste was inevitable. As enormous amounts of time and money were continually poured into the means of potential mass annihilation, death became costly and life was cheapened. After all, what assurance was there that nuclear weapons would not someday be used as all other previous means of warfare had been? If everyone was forced to live more and more in the shadow of fear, life could lose its purpose and human beings all sense of confidence in themselves. In such an atmosphere democratic government could easily succumb to authoritarian rule. As all countries sought desperately to insure their own national security, they joined in creating the basis for the most monstrous future insecurity. By resting national security upon the continued exploitation of nature, every country insured the insecurity of all through mutual destruction of the world's physical and biological inheritance. Each government's interest in any kind of conservation of resources was focused on its own national security and defense. Against this emphasis, the concept of a balance of nature appeared highly theoretical and idealistic. Yet, except in the most narrow nationalistic and chronological terms, it was doubtful that such a restricted view of conservation would yield any but the most short-run advantages. If the resources of the globe were exhausted to provide Americans, for example, with what was at best a limited type of temporary military security, it was not difficult to foresee increasing worldwide unrest and universal disaster. Neither the United States nor any nation could expect to go on arming itself at the expense of nature, even if the rest of the world was willing to meet its terms.

Perhaps the United States in leading the Western world to the discovery of nuclear weapons was also leading it to its destruction. If so, fruition in civilization and in methods of warfare had been reached simultaneously. In bringing the process of war to its ultimate efficiency, Western society also exposed the weakness of its civilization. Part of the long-standing ideological conflict between East and West, between Orient and Occident, was a differing conception of nature. In the past the great achievements of the Western world had been founded on the successful exploitation and adaptation of nature to man's needs. But continued success might now require moderation and self-control. In this respect the more primitive and passive ideologies of the East could, in the long run, prove more enduring. An industrial technology offered men material progress, but it also increased the power and range of man's destructive abilities. It followed, therefore, that general survival might depend on the resistance of the rest of the world to the example of the West. Pacifist opponents of war had become in many ways the ultimate realists of the twentieth century. Perhaps the only hope of escape from mankind's dilemma was that, if the awful prospect of nuclear war could not make men angels, it might at least save the world by making them cowards. If disarmament and control of nuclear tests and weapons could not be reached by international agreement, the price of world survival might depend on the unilateral disarmament of one nation more dedicated to peace than to a meaningless victory.

A civilization that elevated power and violence to a way of life could hardly reverence nature or achieve real balance and harmony between man and nature. So long as the popular mind was focused on fear of imminent attack from abroad, it could not concentrate on problems of peace, much less on the question of man in his relationship with nature. Psychologically it made little sense to urge conservation and economy and, at the same time, seek public support for an expanding arms program. Military deterrence was based on a highly simplified view of the cold war and power struggle. At the same time diplomacy, instead of playing down the importance of ideology, degenerated into the art of avoiding any concession or compromise. If nation or nations could not accept the need for stability and balance in the current world, what prospect was there of realizing an atomic truce? Arms races never stood still. If preparedness ever prevented war, it had done so merely for short intervals before the final outcome in a pitched battle. In democracies and dictatorships alike the cold war created popular animosities and tensions that could not easily be diverted to peace and understanding, either of man with man, or between man and his environment.

"As a result," declared General Omar N. Bradley, "we are now speeding toward a day when even the ingenuity of our scientists may be unable to save us from the consequences of a single rash act or a lone reckless hand upon the switch of an uninterceptible missile. For twelve years now we've sought to stave off this ultimate disaster by devising arms which would be both ultimate and disastrous. . . . If we are going to save ourselves from the instruments of our own intellect, we had better soon get ourselves under control and begin making the world safe for living." [8] General Bradley's words illustrated the serious concern over the future of civilization that was constantly being expressed by leaders in a wide range of professional and scientific fields throughout the world. There was no lack of urgent warnings, even though governments continued officially to be secretive and to withhold from the general citizenry some of the grimmer facts of the new methods of mass annihilation. Education in the horror of war, however, was hardly the major problem. Few informed persons questioned any longer the technical and real possibility that another world war could eliminate human life on the planet.

No scientist or military strategist could say that the worst was certain; on the other hand, none could be really optimistic. Following the first atomic explosions a number of scientists left their research activities to explain to the public the tremendous implications for good or evil of the revolution in nuclear physics. The scientists' new-found sense of social consciousness and responsibility to their fellow men was a measure of their anxieties in the face of their unprecedented discoveries. Symbolic of the scientists' concern was their founding of the *Bulletin of the Atomic Scientists* with its figure of the hands of the clock pointing to a few minutes before midnight. Nevertheless, the atomic scientists continued to make bombs, while their colleagues helped to create new means of spreading disease and death through the germs and poisons of biological and chemical warfare. Indeed, so inhuman were the prospects of modern weapons that the average individual lost his capacity to be shocked, and most persons adopted the protective mask of resignation or of withdrawal into their own private interests and problems.

However much one might wish that the scientists had not brought man face to face with the power of collective destruction, society could not abdicate its general responsibilities. Mankind as a whole, and not the scientists alone, had to make what were not technical but major political and social decisions. Few scientists were likely to follow the example of Leonardo da Vinci who was supposed to have destroyed his models of harmful inventions. Modern scientists often had a sublime faith in technology and its ability to solve the social problems it created. Lacking that confidence, society had to seek political solutions. For example, air and sea pollution, especially from radioactive fallout, was an international problem that would require consideration even if bomb testing were stopped. As the authors of a report to Congress on scientific developments and foreign policy pointed out, "In the coming decade, science and technology will provide new means to use the vast resources of the oceans, to exploit the Arctic and Antarctic, to explore space, perhaps to affect climates. Unless better ways of cooperation are established, these advances into new frontiers will intensify international tensions." [9]

Most frightening of all the possibilities inherent in scientific progress was the suggestion that man was losing his power of responsible decision. In the technological complex and political chaos of the neo-modern world, war could come as a matter of miscalculation by a man or machine. War or peace might depend on an accident of automation, with missile or anti-missile sent aloft by mistake. Or someone could press the wrong button and release the massive forces of attack and counterattack constantly poised and ready in all parts of the world. Human error was a necessary part of life, but in the historic past the consequences of a mistake had never loomed so large.

Even if pushbutton war still seemed an exaggerated and remote possibility, the subtle psychological threat to mankind and nature posed by the revolution in science remained. Eugene Rabinowitch, editor of the *Bulletin of the Atomic Scientists*, noted that man throughout history had always been threatened by disasters he could understand and in large part cope with. But the "world in which nuclear forces are on the loose is a world in which man cannot survive by the same kind of endurance, cleverness, and luck which have permitted him to survive in the 'chemical' world of yesterday. The rapid advance of scientific thought has projected mankind into an alien world where temperatures are measured in millions of degrees and pressures in millions of atmospheres. Man can survive in this world of incredible violence only by a similarly spectacular progress in social and political wisdom." [10]

The problem therefore was to extricate man from the dilemma of his own making. Perhaps the real clue to his difficulty, after all, was in his relations with nature. Valuing power and exploitation rather than balance and harmony, he had lost the real meaning of life. In what Walter Millis called the hypertrophy of modern war, man had found the nemesis of power.[11]

Lord Acton's famous phrase, "All power tends to corrupt; absolute power corrupts absolutely," had grown trite from repetition. But, as Joseph Wood Krutch pointed out, though accepted as true of both individuals and government, it was not believed of mankind itself.[12]

Yet Acton's dictum described very well the dilemma of modern man in regard to war and civilization. The absolute power inherent in nuclear weapons had achieved the ultimate corruption of civilization. War had come to the end of the road with nothing to offer but death. The frightening reality, however, was that war, despite its futility, might still carry man and nature with it into oblivion. In its final struggle the hypertrophy of war could also be the death of civilization.

XIV

CONCLUSION

As the world entered the second half of the twentieth century there seemed to be no immediate prospect that the problems of possible nuclear war, diminishing natural resources, or continued overpopulation were closer to solution. Men were still trying to resolve age-old difficulties through a further and more intensive exploitation of their physical environment. At the same time man's impressive achievement in exploring outer space, opening new vistas of science and technology, helped to increase his faith in his ability to conquer nature, even if it did not allay the fears of those who continued to ponder the fate of individuals who might be compelled to live in "a huge insensate robot state." [1]

Proponents of a mechanistic society believed that technology could enforce its own kind of balance, revising alike both nature and human nature. As scientists constantly developed new processes and products modern technocrats were confident that man could be released from his dependence upon the natural world. And it was now even thought probable that the realm of interplanetary space would soon become subject to his command. Nineteenth-century Darwinian concepts of the survival of the fittest were reinterpreted in the twentieth century in terms of an inexorable scientific progress. Thus, in the modern world, the old idea of some sort of balance or harmony of man and nature had become enormously complex.

In contrast to those persons who accepted with equanimity each advance of science or to those others who yearned to return to a mythical Golden Age, many citizens saw some hope of gaining a reasonable balance between man and the forces of nature. Civilization was, of course, clearly incompatible with a world in which there was no development or use of natural resources. Nature itself as well as man was ever changing, while the very concept of balance and harmony implied a state of fluidity and adjustment rather than the status quo. Yet it was surely open to doubt whether modern man was truly civilized if his life and progress invited its own destruction through the ruthless exploitation of the natural world that was his home.

In the United States the question of the relationship of man and nature was highlighted by the fact that no other nation equaled the American people in their paradoxical ability to devastate the natural world and at the same time mourn its passing. Historically the American faith in progress and technology had had to meet the criticisms of the agrarian followers of Jefferson and of individualists like Emerson, Thoreau, and Henry George. The conquest of the frontier was matched by the romantic appreciation of the beauty of the West and by a growing attention to the conservation of its resources. Political philosophies of expansion and war were countered by ideals of individual freedom and peaceful social progress. By the 1960's many of these contrasting historic views of the proper relationship of man and nature seemed to assume a new vitality, and the subject continued to inspire much thoughtful commentary.

The very attention being given to the interrelations of man and nature was in itself recognition of the importance of the concept of harmony and balance. No matter how fantastic the progress of science in enabling man to control his habitat, there remained the nagging question of his power to do so in an environment in which all humanistic, nonmaterial values might be lost. Man himself stood as the greatest challenge to his own survival, whether in the flaming terror of a nuclear war or in the uncontrolled increase of his numbers. Although a nuclear war could make all other anxieties obsolete, the extraordinary modern growth of the world's population was an even more basic factor in upsetting the balance of nature. Even if science solved all remaining shortages of food and resources, it could hardly overcome the social and political pressures resulting from the overpopulation of large areas of the globe. This, in itself, could lead to war, for surely most nations would fight rather than starve. Moreover, the world's ever-growing population threatened to overcome any idea of a natural balance by eliminating the living space of the various forms of life, whether plant or animal. It would indeed be a curious victory for man if he were to be the only species to survive, and if he were to do so in a chemical world devoid of natural plant or animal life.

It is true that the existence of living beings necessitates constant change. Thus the growth of population requires a reduction in the numbers of wild animals. Yet it was also true that man's own evolutionary progress was affected by other animals. In obliterating all lower forms of life, man might be destroying some part of himself. Or he might even be preparing the way for his successor since some insects or animals—the housefly and rat, for example—find man's environment more congenial than their old natural one. While the circumstances of civilization demanded that the benefits of nature be channeled where they could be of the most use to man, it was not just sentimental idealism that led man to try to preserve the varied species of animal life—in parks, if not in their free state. In the words of Heinz Heck, director of the zoological gardens in Munich, "A race which treats its fellow-creatures, 'dumb animals' though they may be, without consideration and in a spirit of hostility is unable, in the long run, to uphold its civilization. It loses its vigour, its spiritual possessions melt away, and its creative strength ebbs more and more the wider the gap becomes which separates it from its fellow beings." [2]

Nowhere were the pressures of population more real than in the great urban centers of the world. More cramped for space than the wild animals in the zoos, even when like the animals they were decently housed and fed, millions of people endured the unnatural life of large cities. "Metropolitan America is in a squeeze," writes a contemporary authority. "The space it uses for living and to make a living has become cluttered to the point of frustration. Efforts to relieve congestion are feeble by comparison with the forces that make it worse." [3]

The average man was most likely to appreciate the wisdom of Sir Francis Bacon's dictum—"We cannot command nature except by obeying her"—only when the forces of nature revealed themselves in some dramatic catastrophe. Floods and storms, lightning and earthquake, were all tangible evidences of the power of nature which man could experience and understand. Confronted from time to time by nature's less pleasant manifestations, society was then somewhat moved to consider the desirability of achieving a harmony and balance with nature in place of more strenuous efforts at control or conquest.

"One of the emptiest of man's many boasts," an editorial writer in the *New York Times* declared, "is that he has mastered his environment, this earth on which he lives. Spring after spring he sees the floods come, sometimes minor, sometimes major and disastrous, but always water overflowing and invading land that man doesn't want flooded." Despite his accomplishments in building dams and dikes and in forecasting more accurately the weather, it was probable that man's own work in devastating woodlands, reclaiming valleys and bogs, and paving great urban areas had increased the potential flood water. "Flood control," the editorial concluded, "is a compromise, always, an attempt to restore old balances. Weather can upset all the balances. . . . It sometimes seems that nature is determined to show man who is in control. Nature has no such purpose, of course, but man might well temper his boasts from time to time. Especially when the floods come." [4]

Even though man did not always understand his own role in disturbing the balance of nature, he was apprehensive lest the disruption of its inner harmonies should menace his own personal health and well-being. Not only fire and flood, but the more gradual and subtle effects of the smog that settled over large cities, the chemicals that adulterated more and more foods, and the nuclear fallout that spread everywhere created a genuine alarm among a wide variety of concerned citizenry. Thus the nature writer and conservationist Rachel Carson in her book *Silent Spring* seemed to touch popular sensitivity with her thesis denouncing the extensive and careless use of insecticides.[5]

On commercial farmland and suburbanites' lawns chemicals

of all sorts and varying degrees of lethal power were scattered promiscuously in the effort to inhibit the growth of undesired insect and plant life. As a part of this process of chemical control some foods were poisoned, fish, birds, and wildlife destroyed, and the health of the farmer and other users endangered. Obviously not all of the chemicals on which modern agriculture had become so heavily dependent were a threat to man or nature. There also had to be some comparison of potentially harmful effects with possible present or future benefits. But it also seemed clear that the widespread use of chemical killers to combat every weed, grub, or bug represented in an extreme form man's frequently malign influence on the natural world.

In taking a fresh look at the economy of nature and the ecology of man, Marston Bates, professor of zoology at the University of Michigan, concluded: "In defying nature, in destroying nature, in building an arrogantly selfish, man-centered, artificial world, I do not see how man can gain peace or freedom or joy. I have faith in man's future, faith in the possibilities latent in the human experiment: but it is faith in man as a part of nature, working with the forces that govern the forests and the seas; faith in man sharing life, not destroying it." [6]

The United States, almost alone among nations, was still a land of plenty, relatively unaffected by scarcities of food or mineral resources. But the tremendous American capacity to produce invited its own problems. The oversupply of both farm and manufactured goods encouraged a national attitude of carelessness and waste. In all the world only the American people could afford to be improvident, heedless of the dire warnings of conservationists. As factory and farm piled up their bounty logic suggested the wisdom of slowing production, trimming the amount of material goods to actual consumptive needs. But fears of a possible economic depression or of wartime shortages encouraged political support for the maintenance of full production. Thus the national stockpile of strategic goods and farm commodities grew, even though much of the accumulation was actually wasteful. An austere national diet, balancing production and consumption, however sensible such a policy might be in terms of the future of life on the planet, ran counter to all popular notions of progress. But, if scarcity was bad, it was also well to remember that an overabundance was not necessarily good.

Although it was difficult for most men to take the long view that the concept of balance and harmony required, it seemed to many authorities that only some more equable adjustment between man and nature could insure the continuance of civilization in anything like its historic form. Plainly either a nuclear war or the maintenance of the twentieth century's high rate of population growth would, in two drastically differing ways, bring about the effective collapse of civilization. Nor could technology of itself insure the future. In comparison with all extreme prescriptions the idea of man living in balance with nature offered the most hopeful course of action for the future. And, at the same time, such a philosophy held out the key to a harmonious, peaceful, and truly civilized world.

NOTES

I. INTRODUCTION

1. Lewis Mumford, *The Brown Decades: A Study of the Arts in America, 1865–1895* (New York: Dover, 2d ed., 1955) p. 59.
2. C. J. Glacken, "Changing Ideas of the Habitable World," in William L. Thomas, Jr., ed., *Man's Role in Changing the Face of the Earth* (Chicago: University of Chicago Press, 1956), pp. 70–88.
3. J. C. Greene, "Objectives and Methods in Intellectual History," *Mississippi Valley Historical Review*, XLIV (June, 1957), 61–62.
4. *This Is Du Pont: The Story of Technology* (Wilmington: Du Pont Company, 1958), p. 1; *This Is Du Pont: The Story of Man and His Work* (Wilmington: Du Pont Company, 1959), p. 32.
5. James C. Malin, "The Contriving Brain as the Pivot of History . . . ," in George L. Anderson, ed., *Issues and Conflicts: Studies in Twentieth Century American Diplomacy* (Lawrence: University of Kansas Press, 1959), p. 359.
6. Joseph Wood Krutch, *The Measure of Man* (Indianapolis: Bobbs-Merrill, 1954), pp. 25–26.

II. THE AGRARIAN DREAM

1. Quoted in Russell Lord, *Behold Our Land* (Boston: Houghton Mifflin, 1938), p. 77.

2. Hoxie N. Fairchild, *The Noble Savage: A Study in Romantic Naturalism* (New York: Columbia University Press, 1928), p. 2; Arthur O. Lovejoy and George Boas, *Primitivism and Related Ideas in Antiquity* (Baltimore: Johns Hopkins Press, 1935), chs. 1–2.

3. Hans Huth, *Nature and the American* (Berkeley: University of California Press, 1957), p. 9.

4. *Ibid.*, pp. 9, 58.

5. *Ibid.*, p. 16.

6. *Writings of Thomas Paine*, ed. Moncure D. Conway (New York: Putnam, 1894–1896), IV, 339. See also Harry H. Clark, *Representative Selections of Thomas Paine* (New York: American Book Company, 1944), pp. xv–xxvii.

7. Quoted in Henry Nash Smith, *Virgin Land: The American West as Symbol and Myth* (Cambridge: Harvard University Press, 1950), p. 125.

8. Quoted in Huth, *Nature and the American*, p. 22.

9. Quoted in Smith, *Virgin Land: The American West as Symbol and Myth*, p. 126.

10. *Ibid.*, pp. 127–28.

11. Ralph N. Miller, "American Naturalism as a Theory of Nature," *William and Mary Quarterly*, 3d ser., XII (January, 1955), 74–95; Brooke Hindle, *The Pursuit of Science in Revolutionary America, 1735–1789* (Chapel Hill: University of North Carolina Press, 1956), p. 322.

12. "Notes on Virginia," Query XIX, *Writings of Thomas Jefferson* (Washington: Thomas Jefferson Memorial Association, 1903), II, 228–30.

13. Jefferson to Madison, December 20, 1787, *ibid.*, VI, 392–93.

14. Jefferson to Thomas Leiper, January 21, 1809, *ibid.*, XII, 238.

15. Letter of March 31, 1809, quoted in William A. Williams, *The Tragedy of American Diplomacy* (Cleveland: World, 1959), p. 23. See also Jefferson to James Jay, April 7, 1809, *Writings* (Memorial ed.), XII, 271.

16. Eugene T. Mudge, *The Social Philosophy of John Taylor of Caroline* (New York: Columbia University Press, 1939), pp. 158 ff.

17. Smith, *Virgin Land: The American West as Symbol and Myth*, p. 260.

18. Henry B. Parkes, *The American Experience* (New York: Knopf, 1947), p. 60.

III. THE ROMANTIC VIEW

1. "A Memorandum of M. Austin's Journey . . . 1796–1797," *American Historical Review,* V (April, 1900), 527, 542.
2. Henry Nash Smith, *Virgin Land: The American West as Symbol and Myth* (Cambridge: Harvard University Press, 1950), pp. 53 ff. See also Arthur K. Moore, *The Frontier Mind: A Cultural Analysis of the Kentucky Frontiersman* (Lexington: University of Kentucky Press, 1957), pp. 186 ff.
3. Perry Miller, "The Romantic Dilemma in American Nationalism and the Concept of Nature," *Harvard Theological Review,* XLVIII (October, 1955), 247.
4. Smith, *Virgin Land: The American West as Symbol and Myth,* pp. 71–72. See also Roy H. Pearce, *The Savages of America: A Study of the Indian and the Idea of Civilization* (Baltimore: Johns Hopkins Press, 1953).
5. "Second Annual Message" (December 6, 1830), in J. D. Richardson, ed., *A Compilation of the Messages and Papers of the Presidents, 1789–1897* (Washington: 1896–1899), II, 521.
6. Alexis de Tocqueville, *Democracy in America,* ed. Phillips Bradley (New York: Knopf, 1945), II, 74.
7. *Ibid.,* I, 290–92.
8. William Charvat, *The Origins of American Critical Thought, 1810–1835* (Philadelphia: University of Pennsylvania Press, 1936), pp. 71 ff., 91–92. See also Joseph W. Beach, *The Concept of Nature in Nineteenth-Century English Poetry* (New York: Macmillan, 1936); Norman Foerster, *Nature in American Literature* (New York: Macmillan, 1923), *passim.*
9. Oliver W. Larkin, *Art and Life in America* (New York: Rinehart, 1949), p. 141.
10. Quoted *ibid.,* p. 143.
11. James Fenimore Cooper, "Home as Found" (1838), in *Works* (Mohawk ed. New York: Putnam, 1896–1897), p. 126.

12. Cooper, "The Pioneers" (1823), *ibid.*, pp. 99–100.

13. Nathaniel Hawthorne, *Main-Street*, ed. Julian Hawthorne (Canton, Pa.: Kingate Press, 1901), p. 16.

14. Hans Huth, *Nature and the American* (Berkeley: University of California Press, 1957), p. 66.

15. Andrew Jackson Downing, *Rural Essays* (New York: Putnam, 1853), p. 144.

16. *Ibid.*, p. 157.

17. Huth, *Nature and the American*, pp. 67–69.

18. Lewis Mumford, *The Brown Decades: A Study of the Arts in America, 1865–1895* (New York: Dover, 2d ed., 1955), pp. 62–63.

19. Smith, *Virgin Land: The American West as Symbol and Myth*, pp. 37 ff.

IV. TECHNOLOGY AND PROGRESS

1. John Quincy Adams, *An Oration . . . on the Fourth of July, 1831* (Boston: 1831), p. 38; "First Annual Message" (December 6, 1825), in J. D. Richardson, ed., *A Compilation of the Messages and Papers of the Presidents, 1789–1897* (Washington: 1896–1899), II, 311.

2. Quoted in Charles and Mary Beard, *The American Spirit* (New York: Macmillan, 1942), pp. 217–18.

3. *Ibid.*, p. 218.

4. Henry C. Carey, *The Past, the Present, and the Future* (Philadelphia: 1848), p. 416.

5. Henry C. Carey, *The Harmony of Interests, Agricultural, Manufacturing, and Commercial* (New York: 1856), p. 229.

6. *Ibid.*, pp. 228–29.

7. Arthur A. Ekirch, *The Idea of Progress in America, 1815–1860* (New York: Columbia University Press, 1944), pp. 116–17; H. A. Meier, "Technology and Democracy, 1800–1860," *Mississippi Valley Historical Review*, XLIII (March, 1957), 618–40; Bernard Bowron, *et al.*, "Literature and Covert Culture," *American Quarterly*, IX (Winter, 1957), 377–86.

8. Robert Dale Owen, *Wealth and Misery* (New York: 1830), p. 14.

9. *Working Man's Advocate*, I (February 6, 13, 1830), 1, 1.
10. Charles Lane, "Life in the Woods," *The Dial*, IV (April, 1844), 415–25.
11. Albert Brisbane, "Exposition of Views and Principles," *The Phalanx*, I (October 5, 1843), 7.
12. Horace Greeley, *The Crystal Palace and Its Lessons* (New York: 1851), p. 26.
13. Zoltan Haraszti, *The Idyll of Brook Farm* (Boston: Public Library, 1937), p. 13.
14. Quoted in Octavius B. Frothingham, *George Ripley* (Boston: 1882), pp. 172–73.
15. *Scientific American*, II (June 19, 1847), 309.
16. Quoted in Walt Whitman, *I Sit and Look Out: Editorials from the Brooklyn Daily Times*, ed. Emory Holloway and Vernolian Schwarz (New York: Columbia University Press, 1932), p. 133.
17. Ekirch, *The Idea of Progress in America, 1815–1860*, pp. 120 ff.

V. TRANSCENDENTAL HARMONY: EMERSON

1. Ralph Rusk, *The Life of Ralph Waldo Emerson* (New York: Scribner, 1949), pp. 243–44.
2. *The Early Lectures of Ralph Waldo Emerson*, ed. Stephen E. Whicher and Robert E. Spiller (Cambridge: Harvard University Press, 1959), p. xvii. See also Sherman Paul, *Emerson's Angle of Vision: Man and Nature in American Experience* (Cambridge: Harvard University Press, 1952), ch. 3.
3. *Early Lectures*, p. 10.
4. *Ibid.*, pp. 14, 19–20.
5. *Ibid.*, pp. 20 ff.
6. *Ibid.*, pp. 34, 39, 43–44, 49.
7. "Self Culture" (1838), *The Works of William E. Channing* (Boston: American Unitarian Association, 1888), p. 19.
8. "Nature" (1836), *Complete Works of Ralph Waldo Emerson* (Centenary ed., Boston: Houghton Mifflin, 1903–1906), I, 4.

9. *Journals of Ralph Waldo Emerson*, ed. Edward W. Emerson and Waldo E. Forbes (Boston: Houghton Mifflin, 1909–1914), III, 207–8.

10. *Ibid.*, IV, 118–19.

11. *Early Lectures*, p. 24; *Journals*, III, 372, 326–27.

12. *Journals*, IV, 473, 14–15; V, 58.

13. *Ibid.*, I, 299–301.

14. "Nature," *Complete Works*, I, 61.

15. "Self-Reliance" (1841), *Complete Works*, II, 84.

16. Quoted from various essays in *Complete Works*, II, 90, 286, 122, 302 ff.

17. *Journals*, V, 285; VI, 14–15.

18. Vernon L. Parrington, *Main Currents in American Thought* (New York: Harcourt, Brace, 1927–1930), II, 386–99.

19. *Journals*, VI, 397.

20. Herbert W. Schneider, *A History of American Philosophy* (New York: Columbia University Press, 1946), p. 286.

VI. TRANSCENDENTAL HARMONY: THOREAU

1. "Walden" (1854), *Writings of Henry David Thoreau* (Manuscript ed., Boston: Houghton Mifflin, 1906), II, 143, 153.

2. Sherman Paul, *The Shores of America: Thoreau's Inward Exploration* (Urbana: University of Illinois Press, 1958), p. 196.

3. Quoted *ibid.*, p. 182.

4. "Natural History of Massachusetts" (1842), *Writings*, V, 105.

5. "Walden," *Writings*, II, 12–13.

6. *Ibid.*, II, 34, 37, 44, 41.

7. *Ibid.*, II, 15, 102.

8. "Paradise (to be) Regained" (1843), *Writings*, IV, 302, 284.

9. *Journal* (Vols. I–XIV; in *Writings*, Manuscript ed., Vols. VII–XX; cited herein separately from *Writings*), I, 368.

10. "Walden," *Writings*, II, 101, 103, 227.

11. *Ibid.*, II, 232; *Journal*, I, 341; IV, 445.

12. *Journal*, I, 253, 444; IX, 202.

13. "Walden," *Writings*, II, 356, 320.
14. *Ibid.*, II, 349–50.
15. *Journal*, IV, 323–25; XII, 387.
16. *Ibid.*, II, 52, 451.
17. "Walking" (1862), *Writings*, V, 205–7, 212, 216; "Life without Principle" (1863), *ibid.*, IV, 457.
18. *Journal*, VIII, 7–8; XI, 78–79; XIV, 295.
19. *Ibid.*, VIII, 220–21; IX, 205, 208–9.
20. "Civil Disobedience" (1849), *Writings*, IV, 369; "Walden," *ibid.*, II, 355, 358–59.
21. Paul, *The Shores of America: Thoreau's Inward Exploration*, pp. 245–46.

VII. GEORGE PERKINS MARSH: PIONEER

1. David Lowenthal, *George Perkins Marsh, Versatile Vermonter* (New York: Columbia University Press, 1958), *passim.*
2. George P. Marsh, *The Goths in New-England* (Middlebury: 1843), *passim.*
3. George P. Marsh, *The American Historical School* (Troy: 1847), pp. 15–16.
4. George P. Marsh, *Address Delivered before the Agricultural Society of Rutland County* (Rutland: 1848), p. 3.
5. *Ibid.*, pp. 4–6.
6. *Ibid.*, pp. 17–18.
7. George P. Marsh, *Man and Nature; or, Physical Geography as Modified by Human Action* (New York: Scribner, 1864), Preface.
8. *Ibid.*, p. 34; ch. 1, *passim.*
9. *Ibid.*, pp. 38, 44.
10. *Ibid.*, p. 26.
11. *Ibid.*, pp. 336–38, 351.
12. *Ibid.*, pp. 179, 328, 35, 549.
13. George P. Marsh, "Irrigation: Its Evils, the Remedies, and the Compensations," *Report of the Commissioner of Agriculture for the Year 1874* (Washington: 1875), pp. 363 ff.
14. *Ibid.*, p. 373.

15. *Ibid.*, pp. 373–76.
16. William L. Thomas, Jr., ed., *Man's Role in Changing the Face of the Earth* (Chicago: University of Chicago Press, 1956).

VIII. CONSERVATIONIST IDEOLOGY

1. Hans Huth, *Nature and the American* (Berkeley: University of California Press, 1957), pp. 144, 151–52.
2. *Ibid.*, pp. 103–4, 152 ff.
3. *Ibid.*, p. 159; A. Hunter Dupree, *Science in the Federal Government* (Cambridge: Harvard University Press, 1957), pp. 199 ff.
4. Powell to the Secretary of the Interior, April 24, 1874, in *Geographical and Geological Surveys West of the Mississippi* (43 Cong., 1 Sess., House Report 612, May 26, 1874), p. 10.
5. John W. Powell, *Report on the Lands of the Arid Region of the United States* (45 Cong., 2 Sess., House Ex. Doc. 73, 1878), *passim.*
6. *Bulletin of the Philosophical Society of Washington*, Vol. VI (Washington, 1884), pp. li–lii.
7. George P. Marsh, *Man and Nature* (New York: Scribner, 1864), p. 232.
8. H. T. Pinkett, "Gifford Pinchot and the Early Conservation Movement in the United States" (MS. Ph.D. dissertation, American University, 1953), pp. 8 ff.
9. *Ibid.*, pp. 11a ff.; *U.S. Statutes at Large*, XXX, 35.
10. Erich W. Zimmermann, *World Resources and Industries* (New York: Harper, 1933), p. 785. See also Samuel P. Hays, *Conservation and the Gospel of Efficiency* (Cambridge: Harvard University Press, 1959), pp. 40 ff.
11. J. L. Bates, "Fulfilling American Democracy: The Conservation Movement, 1907 to 1921," *Mississippi Valley Historical Review*, XLIV (June, 1957), 31.
12. Hays, *Conservation and the Gospel of Efficiency*, pp. 35, 73, 265–66.
13. Gifford Pinchot, *The Fight for Conservation* (New York: Doubleday Page, 1910), pp. 4, 20.

14. Gifford Pinchot, *Breaking New Ground* (New York: Harcourt, Brace, 1947), p. 49.

15. Pinkett, "Gifford Pinchot and the Early Conservation Movement in the United States," pp. 46–47.

16. Huth, *Nature and the American*, pp. 179 ff.; Hays, *Conservation and the Gospel of Efficiency*, pp. 142–45.

17. Pinkett, "Gifford Pinchot and the Early Conservation Movement in the United States," p. 46; ch. 8, *passim.*

18. *Ibid.*, p. 136; ch. 6, *passim.*

19. Hays, *Conservation and the Gospel of Efficiency*, p. 47; Dupree, *Science in the Federal Government*, pp. 245, 249.

20. Pinkett, "Gifford Pinchot and the Early Conservation Movement in the United States," pp. 82, 97–100; *U.S. Statutes at Large*, XXXV, 259.

21. Pinchot, *Fight for Conservation*, p. 117; Pinchot, *Breaking New Ground*, p. 411.

22. E. Louise Peffer, *The Closing of the Public Domain* (Stanford: Stanford University Press, 1951), pp. 69–70, ch. 2; Roy M. Robbins, *Our Landed Heritage* (Princeton: Princeton University Press, 1942), ch. 20.

23. *Preliminary Report of the Inland Waterways Commission* (60th Cong., 1 Sess., Senate Doc. 325, 1908), p. 15; Pinchot, *Breaking New Ground*, p. 25; W. R. Cross, "W J McGee and the Idea of Conservation," *The Historian*, XV (Spring, 1953), 148–62.

24. Pinchot, *Breaking New Ground*, pp. 322–26.

25. Hays, *Conservation and the Gospel of Efficiency*, p. 135; Pinkett, "Gifford Pinchot and the Early Conservation Movement in the United States," pp. 136–37.

26. Pinkett, "Gifford Pinchot and the Early Conservation Movement in the United States," p. 127; *Proceedings of a Conference of Governors* (Washington: 1909), I, 6–12.

27. *Proceedings of a Conference of Governors*, I, 168–71, 333–34.

28. *Ibid.*, I, 173–79.

29. Bates, "Fulfilling American Democracy: The Conservation Movement, 1907 to 1921," *Mississippi Valley Historical Review*, XLIV (June, 1957), 35–36.

30. Hays, *Conservation and the Gospel of Efficiency*, p. 123; Pinchot, *Fight for Conservation*, p. 79.

31. Charles R. Van Hise, *The Conservation of Natural Resources in the United States* (New York: Macmillan, 1910), p. 378.

32. Nathaniel S. Shaler, *Man and the Earth* (New York: Duffield, 1910), p. 1.

IX. A PLANNED SOCIETY

1. Ralph H. Gabriel, *The Course of American Democratic Thought* (New York: Ronald, 1940), p. 204.

2. Lester F. Ward, *The Psychic Factors of Civilization* (Boston: Ginn, 1893), p. 323; ch. 33, *passim.*

3. Simon N. Patten, *Heredity and Social Progress* (New York: Macmillan, 1903), p. 4.

4. Simon N. Patten, *The New Basis of Civilization* (New York: Macmillan, 1907), pp. 25–26. See also his *The Theory of Prosperity* (New York: Macmillan, 1902), *passim.*

5. Charles and Mary Beard, *The American Spirit* (New York: Macmillan, 1942), p. 604.

6. Nicholas Murray Butler, "A Planless World," ch. 2 in the symposium on planning, *America Faces the Future,* ed. Charles Beard (Boston: Houghton Mifflin, 1932), p. 16.

7. Louis L. Lorwin, *Time for Planning* (New York: Harper, 1945), p. 137.

8. *The Public Papers and Addresses of Franklin D. Roosevelt,* (New York: Random House, 1938), I, 627, 632, 646, 752.

9. Henry Wallace, *New Frontiers* (New York: Reynal & Hitchcock, 1934), p. 21.

10. *Ibid.*, p. 21.

11. *Ibid.*, pp. 28–29, 32, 128.

12. *Ibid.*, pp. 274, 286.

13. Pieter W. Fosburgh, *The Natural Thing: The Land and Its Citizens* (New York: Macmillan, 1959), pp. 19–20; Paul K. Conkin, *Tomorrow a New World: The New Deal Community Program* (Ithaca: Cornell University Press, 1959), chs. 1–3.

14. Conkin, *Tomorrow A New World*, ch. 4, and pp. 115, 129–30, 143.

15. *Ibid.*, p. 6; ch. 7. See also Grant McConnell, *The Decline of Agrarian Democracy* (Berkeley: University of California Press, 1953), *passim.*

16. David Lilienthal, *T V A: Democracy on the March* (New York: Harper, 1944), ch. 14; R. G. Tugwell and E. C. Banfield, "Grass Roots Democracy—Myth or Reality," *Public Administration Review*, X (Winter, 1950), 47–55.

17. Broadus Mitchell, *Depression Decade* (New York: Rinehart, 1947), p. 357.

X. RESOURCES AND ENERGY

1. Henry Adams, "A Letter to American Teachers of History" (1910), in *The Degradation of the Democratic Dogma* (New York: Macmillan, 1919), p. 261.

2. J. L. Bates, "Fulfilling American Democracy: The Conservation Movement, 1907 to 1921," *Mississippi Valley Historical Review*, XLIV (June, 1957), 49–50.

3. Richard T. Ely, *et al.*, *The Foundations of National Prosperity* (New York: Macmillan, 1918), pp. 13, 46.

4. E. Louise Peffer, *The Closing of the Public Domain* (Stanford: Stanford University Press, 1951), ch. 7; Roy M. Robbins, *Our Landed Heritage* (Princeton: Princeton University Press, 1942), p. 394.

5. John Ise, *The United States Oil Policy* (New Haven: Yale University Press, 1926), pp. 496 ff.; Erich W. Zimmermann, *World Resources and Industries* (New York: Harper, 1933), ch. 39.

6. Zimmermann, *World Resources and Industries*, p. 784. See also Robbins, *Our Landed Heritage*, p. 401; Peffer, *The Closing of the Public Domain*, ch. 11.

7. Peffer, *The Closing of the Public Domain*, ch. 12.

8. Quoted in A. Hunter Dupree, *Science in the Federal Government* (Cambridge: Harvard University Press, 1957), p. 364.

9. *Ibid.*, pp. 354 ff.

10. Tom Dale and V. G. Carter, *Topsoil and Civilization* (Norman: University of Oklahoma Press, 1955), p. 252.

11. *Ibid.*, pp. vii, 6–7. See also Fairfield Osborn, *Our Plundered Planet* (Boston: Little, Brown, 1948).

12. Dale and Carter, *Topsoil and Civilization*, pp. 35–38.

13. F. M. Heichelheim, "Effects of Classical Antiquity on the Land," in W. L. Thomas, Jr., ed., *Man's Role in Changing the Face of the Earth* (Chicago: University of Chicago Press, 1956), pp. 165–80.

14. Edward Higbee, *The American Oasis: The Land and Its Uses* (New York: Knopf, 1957), p. 35.

15. William Vogt, *Road to Survival* (New York: Sloane, 1948), p. 98.

16. Higbee, *The American Oasis: The Land and Its Uses*, p. 72.

17. Lester E. Klimm and Luna B. Leopold in Thomas, ed., *Man's Role in Changing the Face of the Earth*, pp. 522 ff., 539, 639 ff.

18. W. B. Langbein, "Water Yield and Reservoir Storage in the United States," *Geological Survey Circular 409* (Washington: 1959), pp. 4–5; Luna B. Leopold, "Probability Analysis Applied to a Water-Supply Problem," *Geological Survey Circular 410* (Washington: 1959), *passim.*

19. Andrew H. Clark in Thomas, ed., *Man's Role in Changing the Face of the Earth*, p. 756. See also *ibid.*, pp. 677 ff., 763 ff., 778 ff.

20. James C. Malin, *ibid.*, pp. 350 ff.; O. C. Stewart, *ibid.*, pp. 126–28.

21. W. A. Albrecht, *ibid.*, p. 922.

22. Karl A. Wittfogel, "The Hydraulic Civilizations," *ibid.*, p. 161.

23. Harold E. Thomas, "Changes in Quantities and Qualities of Ground and Surface Waters," *ibid.*, pp. 542 ff.; C. W. Thornwaite, "Modification of Rural Microclimates," *ibid.*, pp. 567–83.

24. Luna B. Leopold, "Water and the Conservation Movement," *Geological Survey Circular 402* (Washington: 1958), p. 6.

25. Thomas, ed., *Man's Role in Changing the Face of the Earth*, pp. 379–80, 448, 851 ff.

26. S. H. Ordway, "Possible Limits of Raw-Material Consumption," *ibid.*, pp. 988, 1008.

27. President's Materials Policy Commission, *Resources for Freedom* (Washington: 1952), pp. 4–5, 140–41.

28. Harrison Brown, *et al., The Next Hundred Years* (New York: Viking, 1957), pp. 18–19.

29. *Ibid.*, pp. 52, 148–51; Brown, "Technological Denudation," in Thomas, ed., *Man's Role in Changing the Face of the Earth*, p. 1030; Harrison Brown, *The Challenge of the Future* (New York: Viking, 1954), p. 229.

30. Quoted in the *New York Times*, December 31, 1956, p. 6.

31. Ralph Lapp, *Atoms and People* (New York: Harper, 1956), p. 286.

XI. POPULATION PROBLEMS

1. H. H. Bartlett, "Fire, Primitive Agriculture, and Grazing," in W. L. Thomas, Jr., ed., *Man's Role in Changing the Face of the Earth* (Chicago: University of Chicago Press, 1956), p. 711.

2. Philip M. Hauser, ed., *Population and World Politics* (Glencoe: Free Press, 1958), p. 10. See also Hauser, *Population Perspectives* (New Brunswick: Rutgers University Press, 1960).

3. Lewis Mumford, *The Culture of Cities* (New York: Harcourt, Brace, 1938), pp. 8, 252. See also Edward Higbee, *The Squeeze: Cities Without Space* (New York: Morrow, 1960).

4. Conway Zirkle, "Benjamin Franklin, Thomas Malthus and the United States Census," *Isis*, XLVIII (March, 1957), 58–62.

5. *Letters and Other Writings of James Madison* (Philadelphia: Lippincott, 1865), III, 101–2, 209 ff., 575 ff.

6. Henry C. Carey, *The Past, the Present, and the Future* (Philadelphia: 1848), p. 92. See also G. J. Cady, "The Early American Reaction to the Theory of Malthus," *Journal of Political Economy*, XXXIX (October, 1931), 601–32.

7. *Yearbook of the United Nations, 1952* (New York: Columbia University Press, 1953), pp. 406, 502–3.

8. Stanford Research Institute, *United States Foreign Policy* (Washington: Government Printing Office, 1959), p. 31.

9. David Riesman, *The Lonely Crowd* (New Haven: Yale University Press, 1950), pp. 6 ff.

10. Stanford Research Institute, *United States Foreign Policy*, pp. 41 ff.

11. Karl Sax, *Standing Room Only: The Challenge of Overpopulation* (Boston: Beacon, 1955), p. 8. See also R. C. Cook, "Malthus' Main Thesis Still Holds," in Resources for the Future, *Perspectives on Conservation* (Baltimore: Johns Hopkins Press, 1958), pp. 72–78.

12. William Vogt, *Road to Survival* (New York: Sloane, 1948), p. 211.

13. *New York Times* (City Edition), Sunday, September 21, 1958, p. 9. Thomas, ed., *Man's Role in Changing the Face of the Earth*, p. 338.

14. Josué de Castro, *Geography of Hunger* (London: Gollancz, 1952), pp. 108 ff.

15. Abel Wolman, "Disposal of Man's Wastes," in Thomas, ed., *Man's Role in Changing the Face of the Earth*, pp. 807 ff.; H. E. Landsberg, "The Climate of Towns," *ibid.*, p. 603.

16. Josué de Castro, *Geography of Hunger*, p. 31.

17. Sax, *Standing Room Only: The Challenge of Overpopulation*, p. 192.

18. Bertrand Russell, *New Hopes for a Changing World* (London: Allen & Unwin, 1951), ch. 5.

19. Sir Charles G. Darwin, "Forecasting the Future," *Engineering and Science*, XIX (April 1956), 22 ff.

20. W. S. Woytinsky, in Hauser, ed., *Population and World Politics*, pp. 46 ff.

21. John D. Durand, in Hauser, ed., *Population and World Politics*, p. 37.

22. Julian Huxley, "Introduction," in Robert C. Cook, *Human Fertility: The Modern Dilemma* (New York: Sloane, 1951), p. viii. See also Frank Lorimer and Frederick Osborn, *Dynamics of Population* (New York: Macmillan, 1934), ch. 9; Fairfield Osborn, "Renewable Resources and Human Populations," in Charles H. Callison, ed., *America's Natural Resources* (New York: Ronald, 1957), p. 15.

XII. THE INDIVIDUAL AND THE MACHINE

1. Siegfried Giedion, *Mechanization Takes Command* (New York: Oxford, 1948), p. v.
2. Arnold Toynbee, *The Prospects of Western Civilization* (New York: Columbia University Press, 1949), p. 23.
3. Susanne Langer, *Philosophy in a New Key* (Cambridge: Harvard University Press, 1951), p. 278.
4. Quoted in Herbert W. Schneider, *A History of American Philosophy* (New York: Columbia University Press, 1946), p. 195.
5. *Overland Monthly*, I (October, 1868), 297–306.
6. Charles A. Barker, *Henry George* (New York: Oxford, 1955), pp. 298, 622.
7. John Crowe Ransom, *et al., I'll Take My Stand: The South and the Agrarian Tradition* (New York: Harper, 1930), pp. ix–xv.
8. Lewis Mumford, *Technics and Civilization* (New York: Harcourt, Brace, 1934), pp. 26–27, 37, 62–63, 105, 302.
9. Thorstein Veblen, *The Instinct of Workmanship and the State of the Industrial Arts* (New York: Huebsch, 1918), pp. 314–15.
10. Mumford, *Technics and Civilization*, p. 272.
11. *Ibid.*, pp. 383, 390 ff., 429 ff.
12. Erich Kahler, *Man the Measure: A New Approach to History* (New York: Pantheon, 1943), pp. 609 ff.
13. John K. Galbraith, "How Much Should a Country Consume," in Resources for the Future, *Perspectives on Conservation* (Baltimore: Johns Hopkins Press, 1958), pp. 89–99.
14. Lawrence K. Frank, *Nature and Human Nature* (New Brunswick: Rutgers University Press, 1951), pp. 15–16, 35.
15. Samuel H. Ordway, *Resources and the American Dream: Including a Theory of the Limit of Growth* (New York: Ronald, 1953), pp. 54–55; chs. 4, 6.
16. E. A. Gutkind, "Our World from the Air: Conflict and Adaptation," in W. L. Thomas, Jr., ed., *Man's Role in Changing the Face of the Earth* (Chicago: University of Chicago Press, 1956), pp. 11 ff., 21, 27 ff.

17. Sir Charles G. Darwin, "The Time Scale in Human Affairs," *ibid.*, pp. 963–69.

18. Carl O. Sauer, "The Agency of Man on the Earth," *ibid.*, pp. 57 ff., 66.

19. Roderick Seidenberg, "The Dominance of Intelligence," *ibid.*, p. 1095.

20. A. E. Burke, "Influence of Man upon Nature—the Russian View: A Case Study," *ibid.*, pp. 1037–38.

21. Guenther Stein, *The World the Dollar Built* (London: Dobson, 1952), p. 91.

22. Raymond B. Fosdick, *Within Our Power: Perspective for a Time of Peril* (New York: Longmans, Green, 1952), pp. 64–65.

23. George Kennan, "How Can the West Recover?" *Harper's Magazine*, CCXVI (March, 1958), 41.

24. R. J. Russell, "Environmental Changes through Forces Independent of Man," in Thomas, ed., *Man's Role in Changing the Face of the Earth*, p. 468.

25. Paul B. Sears, "The Process of Environmental Change by Man," *ibid.*, p. 472.

26. *Ibid.*, pp. 472 ff., 481.

27. Paul B. Sears, "The Steady State: Physical Law and Moral Choice," *The Key Reporter*, XXIV (January, 1959), 8.

28. Paul B. Sears, "Ethics, Aesthetics, and the Balance of Nature," in Resources for the Future, *Perspectives on Conservation*, pp. 106–111. See also Marston Bates, *The Forest and the Sea* (New York: Random House, 1960).

29. Joseph Wood Krutch, *Grand Canyon* (New York: Sloane, 1958), p. 253.

30. *Ibid.*, pp. 241–42.

31. *Ibid.*, pp. 254, 267, 275.

XIII. CIVILIZATION AND WAR

1. Arnold Toynbee, *A Study of History* (London: Oxford, 1954), VII, 449.

2. Charles and Mary Beard, *The American Spirit* (New York: Macmillan, 1942), p. 674.

3. Randolph Bourne, "Unfinished Fragment on the State" (Winter, 1918), in *Untimely Papers* (New York: Huebsch, 1919), pp. 141 ff.; George Santayana, *Dominations and Powers* (New York: Scribner, 1951), p. 79.

4. Harold Laski, *Liberty in the Modern State* (New York: Viking, 1949), p. 20.

5. U.S. Bureau of the Census, *Statistical Abstract of the United States, 1959* (Washington: 1959), Table 307, p. 244.

6. C. Wright Mills, *The Power Elite* (New York: Oxford, 1956), ch. 9.

7. Seymour E. Harris, *The Economics of Mobilization and Inflation* (New York: Norton, 1951), pp. 29 ff., 110 ff.

8. Gen. Omar N. Bradley, "Peaceful Accommodation," *Fellowship*, XXIV (January 1, 1958), 22–23.

9. Stanford Research Institute, *United States Foreign Policy* (Washington: Government Printing Office, 1959), p. 1.

10. Quoted in Ralph Lapp, *Atoms and People* (New York: Harper, 1956), p. 287.

11. Walter Millis, *Arms and Men* (New York: Putnam, 1956), ch. 6.

12. Joseph Wood Krutch, *Human Nature and the Human Condition* (New York: Random House, 1959), p. 139.

XIV. CONCLUSION

1. Fairfield Osborn in Foreword to Edward Higbee, *The Squeeze: Cities Without Space* (New York: Morrow, 1960), p. viii.

2. Heinz Heck, "The Future of Animals," in Wolfgang Engelhardt, *Survival of the Free: The Last Strongholds of Wild Animal Life*, translated from the German by John Combs (New York: Putnam, 1962), p. 13. See also Louis J. and Margery Milne, *The Balance of Nature* (New York: Knopf, 1960).

3. Higbee, *The Squeeze: Cities Without Space*, p. xi.

4. Editorial, *New York Times*, Sunday, April 10, 1960.

5. Rachel Carson, *Silent Spring* (Boston: Houghton Mifflin, 1962).

6. Marston Bates, *The Forest and the Sea: A Look at the Economy of Nature and the Ecology of Man* (New York: Random House, 1960), p. 262.

INDEX